# THE WORKPLACE

*SARS, West Nile Virus and Other Current Issues*

A CLV Special Report

Jamie Knight, Malcolm J. MacKillop and Pamela Leiper

National Library of Canada Cataloguing in Publication

Knight, Jamie
Public health in the workplace / Jamie Knight, Pamela Leiper and Malcolm J. MacKillop.

(CLV special report series)
ISBN 0-459-28302-2

1. Industrial hygiene. 2. Emergency management. 3. Medical emergencies. 4. SARS (Disease). 5. West Nile fever. I. Leiper, Pamela II. MacKillop, Malcolm III. Title. IV. Series.

HD7659.O6K64 2003 658.3'8 C2003-906520-0

THOMSON
CARSWELL

One Corporate Plaza
2075 Kennedy Road
Toronto, Ontario
M1T 3V4

Customer Relations
Toronto 1-416-609-3800
Elsewhere in Canada/U.S. 1-800-387-5164
Fax 1-416-298-5094
World Wide Web: http://www.carswell.com
E-mail: orders@carswell.com

# Foreword

Every person employed in human resource departments in Ontario would benefit from reading this book — not only for effective performance of their duties in workplaces at a time when SARS and West Nile disease have become new threats, but also because it is an excellent general summary of many fundamental aspects of employment law which impact on all who work in organizations at all times.

For others, especially human resource practitioners working in health care institutions or for airlines, the book is a "must read" because these two sectors are in the front lines of the battles against new infectious illnesses. We need the constant co-operation of everyone employed in them if we are to ensure that the global markets for goods, services and travel which have emerged in recent decades do not produce corresponding global exchanges of infectious diseases.

The value of the book is not, however, limited to human resource practitioners. The authors make the important point that "sensible employers should try to take the lead in matters of health and safety" and this means active engagement in these issues at senior executive and board levels. This kind of engagement is as imperative for good governance as it is for good human resource management. The book, consequently, deserves a place on the reading lists of Board Chairs and Chief Executive Officers to help ensure that the organizations for which they are responsible lead in the management of new health care threats rather than simply reacting to events.

In recent weeks, the presence of new threats such as SARS have been described as a new state, demanding that we create a "new normal" set of conditions to deal with them. This view is based on a profound misunderstanding. In fact, the emergence of new infectious diseases has been a normal fact of life for thousands of years. There is really nothing new about it. Any reasonable review of our medical history strongly suggests that this pattern will continue and that anticipation of the problem and sharing of information are the essential themes to effective management of the threat. This book is an important contribution to both of these.

David C. MacKinnon
Executive Director, Ontario Public Health Association

# Preface

The recent SARS crisis has made many of us think about how we should respond to both medical and non-medical emergencies that affect the workplace. Although the extent of the severity of such an event is difficult to predict, we all understand that there are significant health and economic ramifications inherently associated with any crisis of this nature.

The loss of human life and the increased sense of vulnerability that results from a crisis like SARS affects all of us personally and has an impact, either directly or indirectly, on our level of productivity at the workplace.

For employers, medical and non-medical emergencies can have devastating effects on the efficiency and profitability of their enterprise. The recent SARS crisis caught a lot of employers unprepared to deal with the myriad of complicated health, financial and operational issues that resulted from this crisis. There is no doubt that mistakes were made because of the urgency of the matter and the lack of advanced planning for these types of emergencies.

The authors were each engaged in the SARS crisis by providing our clients with the legal and practical guidance required to make the necessary operational decisions. What we noticed was that we were all struggling, both our clients and ourselves, to keep up with the constantly changing status of the crisis and to provide creative and timely advice on an issue with which we had very little prior experience. From this sense of despair and helplessness was born the concept of this book. We were tremendously motivated to create something that could help human resources professionals the next time we are faced with a medical or non-medical emergency that affects the workplace.

Some cases that we handled took many hours of consultation with our clients to resolve. Many of the issues that we faced during the SARS crisis have lead to certain approaches and strategies that we have attempted to share with you in this book. Although every issue that will arise as a result of a medical or non-medical emergency will require a response that is particular to your own workplace, we sincerely hope that we have provided you with a road map within the pages of this book that will prove valuable in the midst of a similar emergency.

The authors would like to recognize the tremendous self-sacrifice and dedication of the many health care workers who risked their own lives and worked many hours in order to battle and defeat the impact of SARS on the rest of the public. Without their dedication and commitment, this medical emergency could have had an even more devastating impact on the health and financial well-being of our society, to an extent not seen in 50 years.

*Jamie Knight, Malcolm MacKillop, Pamela Leiper*

# Dedication

To my wife Betty and our children Geoffrey, Jennifer and Alison. They provide me with inspiration, support and balance, all that I need and more than I deserve.

*Jamie Knight*

To Judy and Evan for their love and support.

*Malcolm J. MacKillop*

I would like to thank Jamie Knight and Malcolm MacKillop for this great opportunity, I am fortunate to have them both as mentors. I also want to thank my parents Ron and Moyra and my partner Maurice who are my biggest supporters and who encourage me to take on new challenges. Finally, I dedicate this book to the memory of my best friend Watson.

*Pamela Leiper*

# Contents

INTRODUCTION — THE MORAL AND LEGAL RESPONSIBILITY OF EMPLOYERS ........ 1

CHAPTER 1 — SARS IN ONTARIO ........ 5

CHAPTER 2 — OBLIGATIONS ON EMPLOYERS ........ 9
Legal Obligations ........ 9
Moral Obligations ........ 10

CHAPTER 3 — APPLICATION OF FUNDAMENTAL EMPLOYMENT LEGISLATION ........ 13
Health and Safety ........ 13
Human Rights ........ 24
Employment Standards ........ 28
Workers' Compensation ........ 31
Privacy ........ 33

CHAPTER 4 — QUASI-CRIMINAL ISSUES ........ 37
Quarantine ........ 37
Obligation to Report ........ 38
Enforcement ........ 38
Detention ........ 39
Rules for Employers ........ 39

CHAPTER 5 — STATUTORY ISSUES SPECIFIC TO SARS ........ 41
SARS Emergency Leave ........ 41
Employment Insurance ........ 43

CHAPTER 6 — HUMAN RESOURCES ISSUES ........ 47
Communication ........ 47
Managing Absenteeism ........ 49
Dealing with Return to Work ........ 59

CHAPTER 7 — COMPENSATION ISSUES .................................... 61

CHAPTER 8 — CONTINGENCY PLANNING FOR EMERGENCIES .................................... 65

CHAPTER 9 — DEALING WITH WORKING FROM HOME......... 69
Benefits of Working from Home .................................... 69
Working from Home Policies .................................... 70
Liabilities Associated with Working from Home .................................... 73

CHAPTER 10 — TRAVEL POLICIES .................................... 75
Liability Issues .................................... 76

CHAPTER 11 — ISSUES SPECIFIC TO AIR TRAVEL WORKERS .................................... 79

CHAPTER 12 — ISSUES SPECIFIC TO HEALTH CARE WORKERS .................................... 81

CHAPTER 13 — OTHER HEALTH RELATED ISSUES IN THE WORKPLACE .................................... 87
West Nile Virus .................................... 87
Second-Hand Smoke in the Workplace .................................... 91
Chemical and Environmental Sensitivities in the Workplace ............. 93
Power Failures .................................... 96

CONCLUSION — REDUCING RISK IN A COMPLEX WORLD ... 99

◆

# INTRODUCTION — THE MORAL AND LEGAL RESPONSIBILITY OF EMPLOYERS

Medical emergencies in the workplace are not a new phenomenon. Health and safety legislation and workers' compensation legislation have been in place for decades to provide guidance to employers regarding injuries, illnesses and avoiding workplace accidents, and to compensate affected workers. Most employers also have experience in accommodating employees with non-work related illnesses and injuries in the workplace, and have some knowledge of the human rights considerations in such situations.

However, most employers have not had experience dealing with communicable diseases in the workplace. With the exception of medical professionals in a health-care setting, dealing with a communicable disease is not commonplace in most workplaces. For the most part, experience with a communicable disease has been limited to isolated situations where an employer has been advised that a particular employee or employee's family member has been diagnosed with an illness such as meningitis, hepatitis or HIV/AIDS. Not only have such occurrences been isolated, but they are also typically situations where the medical community knew what the disease was, as well as the possible treatment options, and how to contain its transmission to other individuals.

In the spring of 2003, this protective glass ball was shattered. The reality of the global world that we live in hit employers around the world, and in Southeast Asia and Ontario in particular. Severe Acute Respiratory Syndrome ("SARS") arrived in Canada in March 2003. SARS was new, unknown, and a highly contagious disease that was being spread around the world by global travel. Ontario Premier Ernie Eves declared SARS a provincial emergency on March 26, 2003. The Premier lifted the provincial emergency effective May 17, 2003. However, the Ministry of Health and Long-Term Care (the "Ministry") has indicated that the reality is that we are now living with SARS in our community and in our health care system. Accordingly, new enhanced practices need to be developed and maintained. The same applies to human resource practices.

This book exists because we consider SARS to be a wake up call, not a "once-in-a-lifetime" fluke. SARS has been tragic, and it has been expensive; however, in the overall scheme of things, it has been far less than catastrophic, at least to date. As of September 3, 2003 there were 438 cases of SARS reported in Canada, and a total of 43 deaths.[1] The World Health Organization reports that between November 1, 2002 and August 7, 2003 there were 8422 cases of SARS and 916 deaths worldwide.[2] These numbers do not take into consideration the number of people who were quarantined as a precaution. If SARS, or something like it, hits Canada again, we should be able to do much better. If something much worse than SARS hits us, perhaps the lessons that we have learned from SARS will help us avert catastrophe, even if the threat is much greater.

This book is meant for employers and their representatives, as well as for employees, including unionized employees, and their representatives. The authors are employment lawyers, typically acting on behalf of management clients. We had to deal with our clients in respect of a variety of issues that were different from what we were used to seeing, in the midst of a situation that was rife with uncertainty. As days turned into weeks, and as one crisis turned into a second one, we became convinced that employers have a fundamental role to play in society.

This role is not just a legal one, although we will certainly examine the many laws that can come into play. This role is also a moral one. Employers have the power and the opportunity to be positive, pro-active forces in society when it comes time to deal with a public health crisis that affects society as a whole. With this book, we want readers to look beyond the strict legal requirements, we want to promote discussion within each workplace as to what makes sense, what is effective, and what balances the interests of the employer and the employees, bearing in mind the overall purpose of the organization and its role in society at large.

This book might well apply to any emergency situation that faces society. For the most part, we have limited ourselves to public health issues. Furthermore, and again for the most part, we have focused on SARS. We may never see SARS again, or we may see it every year, at least for the foreseeable future. For the purposes of this book, SARS is an excellent example of what could happen and what did happen. In the authors' view, SARS is full of lessons. A key objective of this book is to try to identify those lessons so that we all can learn from them.

The foremost lesson from the 2003 SARS outbreak is that businesses were not prepared to deal with a large-scale medical emergency. It is

1 http://www.hc-sc.gc.ca/pphb-dgspsp/sars-sras/cn-cc/20030903_e.html [accessed October 9, 2003].

2 http://www.who.int/en/ [accessed October 9, 2003].

therefore critical, in our view, that businesses review current policies and procedures already in place to ensure that they are prepared for the next medical emergency. The purpose of this book is to review the various legal and moral obligations of an employer and to discuss the many employment issues that can be affected by medical emergencies. We hope to encourage organizations — with employees and management working together — to review what worked and what should be fixed.

We trust that the following pages will provide guidance for all types of workplaces on how to be better prepared next time. As a final note, although this book is written with reference to Ontario legislation, it should be of general application in all Canadian jurisdictions, with the necessary adjustments to deal with legislation that is different in the specifics, but mostly similar in terms of most of the significant concepts that are of relevance to this topic.

> uncontrollable hemorrhaging that filled the lungs, and patients would drown in their own body fluids.
>
> Not only was [this] Flu strikingly virulent, but it displayed an unusual preference in its choice of victims---tending to select young healthy adults over those with weakened immune systems, as in the very young, the very old, and the infirm. The normal age distribution for flu mortality was completely reversed, and had the effect of gouging from society's infrastructure the bulk of those responsible for its day to day maintenance. No wonder people thought the social order was breaking down. It very nearly did.[2]

The above citation summarizes the Spanish Flu pandemic of 1918, which closed out the horrors of the Great War that ravaged Europe from 1914 to 1918. The Spanish Flu killed 40,000 to 50,000 people across Canada.[3] By contrast, in the spring of 2003, although thousands were quarantined as a precaution in Canada, SARS infected only hundreds of people in Canada, killed only tens of people and was found, for the most part, only in health care settings.

Despite the relatively small numbers of Canadians directly affected by SARS, public officials, the health system, and certainly most employers were not prepared for such a widespread and unexpected public health emergency. Whether they are called alarmists or realists, knowledgeable commentators will point out that a flu virus that is airborne means the risk of spread of the virus throughout the community is much greater. Certainly, a flu pandemic has the potential to have an impact on society and the ability to operate a business that would make SARS seem like the good old days.

2 Leonard Crane, "The 1918 Spanish Flu Pandemic and the Honk Kong Incident", http://www.ninthday.com/spanish_flu.htm [accessed October 9, 2003].

3 Jonathan Gatehouse, "SARS: Fear and Loathing of Toronto", *Maclean's*, May 5, 2003, page 21.

◆

# CHAPTER 2 — OBLIGATIONS ON EMPLOYERS

Employment law, which is the body of common law and statutes that impact on the workplace, is a particularly interesting area of the law. At its best, it is dynamic, it is responsive to changes in society, it is part of the fabric, but never should be mistaken for the entire cloth. Successful human resources professionals, including employment lawyers, know that they should not take action or refrain from acting simply because they are within their legal rights to do so. Rather, human resources professionals should act in a fashion that is in the best interests of the organization, taking into account the legitimate interests of the employees, and always in a manner that is lawful. Put another way, human resources professionals should promote the interests of the organization in a manner that is reasonable and respectful towards the employees and consistent with the law. The law sets the parameters around the actions that should be taken. For the most part, perhaps human rights law aside, the law should not be the central force that drives human resources decision-making.

In this respect, as we begin the discussion about the obligations that an employer owes to its employees in dealing with public health issues, certainly we intend to cover the legal obligations. We also mean to raise our views about an employer's moral obligations and the notion of public duty for corporations and other employer organizations. These latter obligations may not be required by law, but they are consistent with the law and with the expectations of society. Moreover, they are reasonable obligations, respectful towards the employees of an organization, and in keeping with the organization's best interests.

## Legal Obligations

To start, and at the very least, an employer must understand its legal obligations when dealing with any medical emergency in the workplace. The legal obligations arise from various statutes that at times may appear to be conflicting. On the one hand an employer has obligations under the applicable health and safety legislation to ensure a safe workplace. On the other hand, these obligations must be met without violating obligations under human rights and privacy legislation. Obligations such as accommodating a disability or a duty not to disclose personal medical information about an employee at times may seem to conflict with the duty to ensure the workplace is safe, particularly when dealing with an employee who has a communicable disease.

There are also legal obligations arising out of employment standards legislation dealing with the absence of employees as a result of medical issues. In the case of SARS, Ontario even enacted emergency legislation to protect the rights of workers affected by SARS. Other legal liabilities may arise from workers' compensation legislation or issues relating to compensation of affected employees.

Through this introduction, and as we will see in much more detail in the following chapter, there are many legal obligations imposed upon employers that could apply to medical issues in the workplace, depending on the factual circumstances of each situation. Employers have a duty to understand their obligations and be prepared to handle such situations in an appropriate manner.

## Moral Obligations

Legal obligations aside, we suggest that employers should embrace their moral responsibilities to the individuals in the workplace and the community at large. Every employer can and should take proactive steps to help prevent employees from contracting various diseases that are currently prevalent. It is in the best interest of every business to stay abreast of what is happening in the community, particularly from a public health standpoint. When a public health issue is raised in the media, employers should dedicate resources to look into the matter. Good employers will want to find out what the public health issue is, who is at risk, what the causes are, what the symptoms are, and what should be done to protect the workforce.

Gaining knowledge for the benefit of employees and, ultimately, for the organization as a whole does not happen by accident. An organization should be structured so that one person, or a group of people, is responsible to keep current with developments in the world at large and how those developments might impact on the workplace. For most workplaces, this is most obviously a human rights function. It is easy for busy people to push this kind of role to the back burner. It takes discipline and real commitment to stay current and to communicate relevant events to the organization and the workforce as a regular part of the job.

The most important role that employers can play is that of a communicator. Advise your employees about any potential health risk, who is at risk, what symptoms should be watched for, and what preventative measures can be taken. Employers to a certain extent have a captive audience and can provide a valuable public service by making employees aware of the danger of whatever disease is prevalent in the community or in the particular workplace. Such communication is a benefit to employees as it may help to keep them healthy. Employers may also directly benefit by avoiding attendance problems in the months to come by letting employees know what to do to protect themselves and their families from contracting diseases such as SARS, West Nile virus and other communicable diseases.

There are steps that can be taken by employers even when there is no immediate public health threat. For example, reminding employees about the benefits of washing their hands at frequent and regular intervals to prevent the spread of disease is always beneficial. In terms of West Nile virus, as another example, taking active steps to prevent and combat mosquito infestation is well within the capability of most workplaces.

Although this suggestion will be counter-intuitive to many employers, one of the most important roles an employer can play is to advise employees to stay home if they have any symptom of a disease or illness. For many this will require a fundamental change in philosophy because of the concern that such encouragement will result in abuse of sick time and increased absenteeism. The concerns about abuse and the fight against excessive absenteeism are important human resources objectives, especially as there are individuals in every organization who abuse sick leave entitlements. It is important to keep in mind that most employees do not have absenteeism problems. Generally speaking, most employees come into work even when they do not feel well. The reasons for employees attending work while ill vary from loyalty and the need to get the job done, to fear of discipline or disapproval if they call in sick.

Unfortunately if this mind-set continues, disease will continue to be spread in the workplace. As such, the impact on the business likely will be greater than if employees are encouraged to stay home for a day or two to recover from an illness rather than coming to work and infecting several previously healthy co-workers.

◆

# CHAPTER 3 — APPLICATION OF FUNDAMENTAL EMPLOYMENT LEGISLATION

## Health and Safety

It is a moral duty, as well as a legal requirement, that employers in all Canadian jurisdictions co-operate with their employees in order to operate safe and healthy workplaces. Not only is that a legal imperative, it makes good business sense and is a key to effective human resources management.

### *Internal Responsibility System for Safe and Healthy Workplaces*

As a matter of law, management and workers are required to co-operate in creating and maintaining a healthy and safe workplace. The internal responsibility system, which is the basis for Canadian health and safety law, places the onus on the workplace participants to create a good working environment. The role of government is to resolve disputes and to ensure that the system is functioning as it should; otherwise, the system mostly relies on the participants in each workplace, and features self-monitoring and self-improvement. Workers are entitled to have input through regular inspections and committee meetings. They are entitled to make recommendations through the minutes of health and safety meetings, which must be considered by the employer. Workers can properly expect management to respond in an effective and timely fashion to reasonable suggestions for improving the workplace environment.

Sensible employers should try to take the lead in matters of health and safety. Although it is a shared responsibility and there is tremendous scope for worker input, it is ultimately the employer and the members of the management team that will bear responsibility in the event of a serious accident or fatality. As a result, it is important for the employer to maintain a comprehensive and current safety policy, to implement procedures for safe operation and dealing with accidents or unsafe incidents, to teach safe practices, and to impose rules against unsafe practices with disciplinary consequences. Consistent with the co-operative nature of health and safety, worker input is very helpful in all of these respects and is probably the best way to raise the profile of health and safety in the workplace. Indeed, effective and ongoing opportunities for input in the development of policies, procedures and rules can result in the employees training themselves to a significant extent.

### *Health and Safety Legislation*

In Ontario, health and safety in the workplace is governed by the *Occupational Health and Safety Act*[1] ("OHSA") and the regulations made pursuant to the OHSA. Part III of the OHSA sets out the duties and liabilities of various stakeholders.

There are specific duties that an employer is responsible for, such as the duty to ensure the safety of the machinery, equipment and processes that are used to carry out the work, and the duty to appoint competent supervisors. However, the paramount duty of the employer is to provide a safe and healthy workplace, and to take every precaution reasonable in the circumstances for the protection of a worker.[2]

In terms of dealing with public health emergencies, it seems clear from health and safety legislation that an employer has a duty to take reasonable precautions to protect employees from contracting contagious diseases, either through the workplace environment or through contact with other employees. When making decisions regarding what is reasonable in the circumstances, an employer should consider the precautions recommended by the Ministry of Health and Long-Term Care, Ministry of Labour, public health officials, medical doctors, researchers and other individuals with knowledge of the particular contagion.

### *Education, Training, Communication and Action*

As we have said, employers should be leaders. Employers have a legislated duty to direct the workers, which includes education and training, as well as the obligation to establish policies and procedures. This duty also requires employers to ensure that employees are familiar with hazards in the workplace and how to protect themselves and other workers. Accordingly, employers should be educating employees about contagions in the workplace. This includes SARS, influenza epidemics and any other contagious diseases that may be present in society and, therefore, a potential unwelcome guest in the workplace. In a similar way, employers should take active steps to prevent undue exposure to other health risks. The West

---

1 R.S.O. 1990, c. O.1, as am.
2 *Ibid.*, s. 25(2)(h).

Nile virus is a good example. Although it is not a contagious disease, there are steps that can be taken to reduce the attractiveness of a work environment to disease-carrying mosquitoes. Employers have to learn about these steps, take them, and encourage all employees to be similarly pro-active.

Our view is that employers should advise their employees of what steps could be taken to protect themselves from contracting West Nile, as well as SARS or other contagions in the workplace. That is good business, good for the society we all live in, and compliant with the letter and spirit of the law. In the case of communicable diseases, employees may request additional personal protective equipment such as masks or gloves. Employers should respond to such requests within reason if there is a threat of exposure to a contagion. A reasoned response is consistent with the obligation to maintain a healthy workplace. When personal protective equipment is provided, an employer must ensure that employees are trained on how to properly use it.

### *Duties of Supervisors*

Supervisors have a duty to ensure that their subordinate employees comply with the OHSA and with safety policies and procedures. This includes the requirement that employees wear protective clothing and use protective devices as may be required. Supervisors should use their disciplinary authority if employees do not work in a safe manner. For example, although we have noted that West Nile virus is not a contagious disease, the provision of personal protective equipment where exposure to mosquitoes in the workplace is likely of great importance.

Mirroring the general duty imposed on companies, supervisors must take every precaution reasonable in the circumstances for the protection of workers. This includes more specific duties, such as advising workers of hazards in the workplace, and providing written instructions to identify and reinforce safe work practices.

### *Duties of Workers*

Workers also have duties under the OHSA related to safety in the workplace. Each worker has an obligation to work safely, to use the available safety devices (and to not render them ineffective), to avoid horseplay, and to report any dangers in the workplace. A worker's obligation to report workplace dangers is important and should be aggressively promoted by employers. The duty to report contagions in the workplace should be considered to be part of the general duty to report dangers in the workplace. If an employee has a communicable disease, it would create a danger in the workplace as, potentially, it could be spread to co-workers.

The legal duty matches the responsible social obligation: if an employee has a communicable disease, it should be reported to management. It is recommended that this duty be specifically spelled out in the workplace safety procedures and policies. Once the employer is aware of the contagion, it can determine what protective measures, if any, need to be taken. Any response to a report about a communicable disease must be sensitive to human rights and privacy concerns. As a practical matter, if the employer is not both sensitive and sensible in dealing with such reports, employees will be effectively discouraged from making them.

### *Public Health Emergencies Are Not New*

Dealing with contagious or potentially hazardous diseases in the workplace is a situation that is not really new to employers. Hospitals and other medical services have always dealt with the realities of infectious disease in the workplace; over time, they have learned to implement progressively better safety precautions. Some employers in non-medical settings have also had to deal with infectious diseases in the workplace, including employees with AIDS/HIV, hepatitis and meningitis to name some of the more obvious situations.

The general duty to ensure that the workplace is reasonably safe requires employers to take reasonable steps to reduce the risk of transmission of known contagions in the workplace. Accordingly, employers have to be in tune with what is happening in the community so that they can communicate potential risks promptly and accurately to employees. If the Ministry of Health or Public Health officials are recommending that certain precautions be taken, then the employer should be implementing appropriate safeguards without delay.

### *Taking Steps to Prevent SARS*

For example, safeguards recommended to prevent SARS include the following:

- Communicate with employees about what SARS is and how to recognize the symptoms
- Advise employees to stay at home if they have any of the recognized symptoms

- Take reasonable measures to ensure that employees at work are not symptomatic
- Deny access to the workplace and send home any employee who exhibits symptoms of SARS or meets the criteria for home isolation established by Ministry of Health or Public Health officials
- Remind employees about the benefits of washing their hands at frequent and regular intervals to prevent the spread of SARS
- Provide antiseptic cleanser in the workplace
- Ensure workplace is kept clean and free of germs to the extent possible
- Ensure ventilation systems in the workplace are clean and in good working order

Although the above safeguards are specific to SARS, they are equally applicable to other contagious diseases, with necessary modifications as advised by Public Health. Moreover, these types of precautions should not be considered temporary during the SARS outbreak. Although SARS appears to be under control, it has not been eradicated. Further, given the global nature of our environment, it is anticipated that there will be other new diseases that we will have to deal with.

The obligations on an employer will vary from industry to industry, as the obligations are based on what is reasonable in the circumstances. The health care industry, travel industry, and to a certain extent the service industry are industries where the obligations may be greater. We will explore issues specific to these particular industries in later chapters.

### *Work Refusals*

Part V of the OHSA gives employees the right to refuse work if they believe the workplace is unsafe. This right applies equally to situations where an employee believes a piece of equipment is unsafe to operate, as well as to a situation where an employee believes he or she may be exposed to a contagion such as SARS. As we will see, the first stage of the refusal is a subjective belief by the employee that safety is at risk, which is a very easy standard for an employee to meet. Even the second stage, which requires the employee to meet an objective standard, is not difficult for an employee to establish, especially in the midst of a public health emergency. In the result, employers could be significantly exposed to work refusals, especially if it becomes known that one or more employees may have attended at the workplace with a communicable disease.

It is important to clarify that the right to refuse work does not apply to certain employees when the danger in question is a normal part of the job or if refusing to perform the work in question would endanger the life, health or safety of another person.[3] In particular, the employees who do not have the right to refuse work include:[4]

- Police Officers
- Firefighters
- Workers at correctional institutions
- Workers at hospitals, nursing homes, homes for the aged, psychiatric institutions, mental health facilities or rehabilitation facility
- Workers at residential group homes or facilities for persons with behavioural or emotional problems, or physical, mental or developmental disability
- Ambulance services or first aid clinics
- Licensed laboratories
- Laundry, food services, power plant or technical service or facility used by one of the above

### *Subjective Right to Refuse Work at the Outset*

At the outset, the right to refuse work is based in the subjective belief of the worker. This subjective belief is a low hurdle and very difficult to challenge. The worker has an initial right to refuse work if he or she believes that:[5]

- the equipment or machinery he or she is required to operate is dangerous to the worker or to other workers,
- the physical condition of the workplace is a danger, or
- there is a contravention of the OHSA or regulations, which creates a danger to the worker or to other workers.

There are two stages to a work refusal. The first stage, following the initial refusal, involves an internal investigation. If the worker is not satisfied with the outcome of the internal investigation, the employer is required to call an inspector from the Ministry of Labour to investigate the complaint.

### *Internal Investigation*

There is no magic here. The worker does not have to make any kind of formal pronouncement — a simple communication to the supervisor will

3 *Ibid.*, s. 43(1).
4 *Ibid.*, s. 43(2).
5 *Ibid.*, s. 43(3).

suffice — that the worker is refusing to carry on with the assigned work because he or she feels it to be unsafe.

Employers should not get carried away too quickly. A supervisor should exercise discretion at this stage if a worker is simply expressing a concern or making a complaint, but is not refusing to carry on with the work. A conservative approach would be to treat the concern or complaint as a refusal until it is resolved, ideally to the satisfaction of the worker. Such a conservative approach may be unduly disruptive to the workplace and is not recommended; it is not necessary and it seems alarmist, especially when dealing with a public health issue. *It is up to the worker to refuse to work.* If the worker does not refuse, then the supervisor should ensure that the work is safe — this is an overriding obligation — but there is no particular reason why the work should not carry forward in the meantime, unless there is an obvious danger. Even if you are conservative in your approach, you should certainly avoid being trapped into encouraging refusals. In dealing with public health issues, employers will want to dampen the fires, not stoke them. Employee concerns should be dealt with in a sensitive and caring manner but, except in extreme cases, employees should be encouraged to continuing working.

Once a refusal to work is clearly identified, the *first stage* is the internal investigation, which must be carried out immediately. The investigation normally would be by the supervisor of the affected area, in the presence of the worker and a worker representative, preferably a certified member of the Joint Occupational Health and Safety Committee. The worker must remain in a safe place near the work area, but should be actively involved in the investigation in order to explain the concern. The worker cannot be assigned other work during this first stage.

It is up to the employer to make a decision at the end of the first stage, either that the work is safe or that modifications must be made or safeguards introduced to remove a potential hazard. Obviously, it is best if the employer can get agreement, at least from the worker representative and ideally from the worker as well.

In the context of an employee refusing to work because of a contagion such as exposure to SARS, the best approach is to discuss the employee's concern and consider whether or not any precautions need to be taken. Perhaps there is personal protective equipment, such as a mask, goggles or other protective clothing that could be provided. Perhaps the medical condition of one or more other employees will have to be assessed, based on an objective assessment of evident symptoms, but having regard to their own privacy. If there is no evidence that an employee had direct contact with an infected person, or there is no evidence that an infected person has accessed the workplace, a work refusal likely will not be successful. However, even where Public Health officials have advised that there is no real risk of transmission of a disease in the community or workplace, because the threshold for a work refusal is much lower at this initial stage employees may still be entitled to refuse work if they have a genuine belief that they are at risk of being exposed to a contagion.

### *Right to Continue to Refuse Work — The Objective Standard*

If the employer decides that the work is safe (with or without some modification), then the supervisor is effectively ordering the worker to resume work. For the worker to rightfully continue with the refusal, the hurdle is now raised to an objective standard. In other words, the worker must now have *reasonable grounds* to believe that the work is likely to pose a danger.

Even with the higher hurdle, it remains difficult from a practical standpoint to contradict the worker at this stage or to treat the continuing refusal as insubordination and subject to discipline.

### *External Investigation*

With a continuing refusal comes the obligation to call in a Ministry of Labour inspector. In the meantime, the work cannot be assigned to another worker, unless the other worker is first informed of the refusal in the presence of the worker representative. The refusing worker may be assigned suitable alternative work if it is available.

The intervention of the inspector represents the *second stage* response to the ongoing refusal. The inspection will be carried out in consultation with the worker, the employer, and the worker representative. It used to be necessary for inspections to be done in the presence of the employer and the worker and worker representative; however, it is now possible for an inspector to conduct a telephone investigation. This is very important, because it could be the preferred method of dealing with public health issues.

The decision of the inspector, which is delivered in writing, usually resolves the matter. Such a decision may be appealed, either by the worker or by the company, in which case an Adjudicator would determine the matter.

Even at the point when workers must meet an objective standard of reasonableness in order to continue to refuse work, it remains difficult to discipline a worker prior to the second stage. However, a refusal that persists after the second stage certainly should result in discipline for insubordination. This discipline may be grieved in a unionized environment or raised in a reprisal complaint. In rare cases, it even could form the basis for a constructive dismissal claim.

### *Reprisals*

Part VI of the OHSA prohibits an employer from taking action against a worker or acting in an intimidating or coercive manner in situations where:

- the worker is acting in compliance with the OHSA or its regulations, including a situation where the worker has invoked the right to refuse work that the worker believes to be unsafe, or
- the worker has given evidence in any proceeding related to issues of safety or health in the workplace.

From a practical standpoint, most reprisal complaints will arise from work refusal situations, especially if the company declares a worker to be insubordinate in the face of an ongoing or recurring refusal to work.

If a worker is a member of a bargaining unit, it is always open to the union to file a grievance on behalf of its member. Even in a non-disciplinary situation and even where there is no particular provision in the collective agreement, an arbitrator now has authority to take jurisdiction of any complaint that involves a labour related statute, like the OHSA.

Separate and apart from the grievance procedure, and available to both union and non-union employees, there is also a right of complaint to the Ontario Labour Relations Board (the "OLRB"). Note that this is an individual right, so that a unionized employee can carry forward a complaint even over the objection of the trade union and a non-union employee can do so as an individual claim.

The OLRB has broad powers of remedy, even where they find that the employer has taken disciplinary action for cause. This latter point is important to understand. The OLRB may find that there is no basis for the reprisal complaint and yet still order compensation, including reinstatement, if it finds that the disciplinary response of the company is unjust. Discipline is difficult to justify in work refusal situations and it can be a sensitive and contested matter, so counsel should be consulted. Although it is understandable that an employer will strive to maintain some order in the workplace, it can be expected that there will be tremendous sympathy for a worker in a situation of a public health emergency, especially if the worker has individual characteristics that increase susceptibility to illness, even if only in the worker's mind.

### *Liability and Penalties under the OHSA*

Offences under the OHSA are strict liability offences. This means that it does not matter if the worker, supervisor or employer did not mean to breach the OHSA or Regulations or for an accident to happen or for a safety hazard to exist. The individual or corporation will generally still be guilty of a strict liability offence unless it can be shown, at the very least, that the individual or corporation acted with due diligence. In other words, most people do not mean for bad things to happen, but that is not the point. The point is whether appropriate and available steps were taken to ensure, as far as possible, that bad things would not happen.

A due diligence defence comes full circle to the reasonable standard discussed earlier; however, do not be lulled into thinking that reasonableness will be easy to establish. Safety is paramount in the eyes of the Ministry of Labour, the OLRB and the Court. To satisfy these third party overseers that it is reasonable, an employer must be able to plainly demonstrate that safety is paramount in its eyes as well.

### *Prosecution*

Both the company and individual employees, especially supervisors, are potentially subject to prosecution. This is particularly true if an accident results in a critical injury or fatality. In public health cases, there is a risk of prosecution in a situation where an employee knowingly brings a contagion into the workplace placing co-workers at risk of serious illness, or where the supervisor or employer knowingly permits a contagion to enter into the workplace without taking reasonable precautions. Some of the most obvious precautions include using simple screening in appropriate cases, and more sophisticated screening if warranted, as well as sending sick people home.

The prosecution charges would be quasi-criminal in nature. The company and any individual charged should certainly consider legal representation, especially as the nature of the legal defences available will have a dramatic impact on potential liability.

A prosecution will proceed very much like a criminal trial. The authority and resources of the Crown, as the prosecuting authority, will be brought to bear on the company or individual charged, who will be in the position of an accused. The goal of the Crown in bringing charges is not simply to establish liability, but to establish guilt. The onus of proof lies with the Crown and is the criminal standard — that is, all of the essential elements of the offence must be established beyond a reasonable doubt. In a public health emergency, there is a clear danger that any prosecution will be a "show-case", carefully watched by the media and the general public, which could result in an unusual application of prosecutorial preparation and zeal.

### *Defences*

Essentially, and depending on the specific nature of the alleged infraction, there are two types of defences to the potential charges. The first defence is the due diligence defence. This is the fundamental defence, the one that should be the bedrock of all your efforts. An employer should be able to show:

- that all reasonable precautions are taken to avoid accidents or the introduction of contagions into the workplace (both generally and with respect to this accident or contagion in particular);
- that all reasonable care to comply with the legislation and with safety directives is taken; and
- that appropriate safety policies and procedures are consistently enforced.

The second defence is the "officially induced error" defence; essentially a defence that places blame back onto civil servants (e.g., "we only did what the inspector or the public health officials told us to do, and that is what caused the mess we are in now"). This is obviously not the best position from which to argue and a company should always strive to be directly responsible for all initiatives in health and safety: accept nothing at face value and question everything.

### *Penalties*

If convicted, an individual is subject to a maximum fine of $25,000 per count or a maximum 12 month jail term, or both. It is important that supervisors and managers are advised of these potential legal consequences so that they can be considered next time they are considering whether or not they should take corrective action when an employee does not use available safety equipment or does not follow established policies or procedures.

The company is subject to a maximum fine of $500,000 per count. "Per count" means that the same set of facts, the same incident, may generate more than one charge, especially if the different offences that are alleged have different elements or involve distinct acts within the same time period. The existence of multiple counts means that there is the possibility of multiple penalties.

## Human Rights

Essentially, human rights law for workplace purposes has two parts: discrimination and harassment. Both of these parts operate to prevent unequal treatment that is based on prohibited grounds.

The protection against discrimination is based on the idea that an employer should recruit, hire and promote without regard to external characteristics — focusing only on the real skills, abilities and qualifications of each person. The protection against harassment is based on the idea that all persons should be free from abuse from co-workers and supervisors that arise from their external characteristics, such as gender, race, cultural background, or disability.

Discrimination can be direct, due to the improper conduct and attitudes of decision-makers. It can also be indirect or constructive in nature, insofar as it arises from barriers to effective recruitment, hiring and promotion practices.

### *Human Rights and Public Health*

In situations involving contagious diseases, an employer has responsibilities to both the infected and the non-infected employees. The Ontario *Human Rights Code*[6] (the "Code") prohibits discrimination and harassment based on disability. In all likelihood, someone with a contagious disease would be considered disabled under the Code. Moreover, the definition of disability under the Code is defined broadly to include the subjective belief that someone is disabled. This means that the Code protects individuals who are the victim of discrimination because an employer or co-worker thinks that person has a disability. So, even if an employee does not have a communicable disease such as SARS, but co-workers discriminate or har-

---

6 R.S.O. 1990, c. H.19 as am.

ass the employee because they think he or she might have the disease, the employer still will be vulnerable to a complaint under the Code.

In some cases there may be an incorrect perception that a disease is associated with a group of people with a certain ethnic background. Since the SARS outbreak that hit Toronto in the spring of 2003 was thought to originate in the Guandong Province of China, the Asian-Canadian community was unjustly stigmatized to a certain extent as a result. Employers have a legal obligation to ensure that a group of employees are not treated differently because of a subjective belief that people of that origin are susceptible to certain diseases and primarily responsible for the spread of the disease. When creating and applying policies regarding SARS or any communicable disease it is important that an employer have reasonable and justifiable grounds for sending an employee home or for denying entry into the workplace. A policy that bans someone from the workplace simply because of the person's race, cultural background or place of origin blatantly violates the Code.

### *Denying Access to the Workplace*

In legitimate circumstances an employer will be justified in denying an employee access to the workplace. Continuing with SARS as our example, such a situation may arise when an employee returns from a trip to a known SARS hotspot, following a period of quarantine, or following time off work to care for a relative who was in isolation or quarantined. However, in order to ensure that these measures are not seen to be discriminatory, it is recommended that any policy should be implemented consistently, regardless of an employee's ethnic background. Further, and rather than arbitrarily banning an employee based on subjective criteria, it is recommended that any at-risk employee should be advised that he or she can return to work if a medical certificate is provided that clears the employee's return.

If an employee refuses to provide a medical certificate when an employer makes a reasonable request for one, or if the medical certificate suggests that an employee is not healthy enough to return to work, then an employer should be able to refuse to allow such an employee back into the workplace without fear of violating the Code.

### *Duty to Accommodate*

Under the Code, an employer has a duty to accommodate a disabled employee to the point of undue hardship. The only factors to be considered when determining undue hardship are cost, sources of outside funding, and health and safety requirements. Therefore, a co-worker's unwillingness to work with an individual will not likely be considered in the determination of undue hardship, particularly when there is no evidence of any risk to the co-worker.

An employer will be required to accommodate the employee to the point of undue hardship even when dealing with an employee who has a contagious disease. Accommodation may include providing the employee with personal protective equipment, providing alternative work, allowing the employee to work from home, or providing the employee time off work, without pay, but without affecting ongoing employment.

### *Avoiding a Poisoned Workplace*

In addition to the duty to accommodate a disabled employee, employers are obliged to ensure that the fear that other employees may have of being infected in the workplace does not result in discrimination or harassment against the person or group of persons whom are feared. As mentioned above, the Code requires employers to provide a workplace free from harassment. The employer cannot allow the workplace to become poisoned with fear. All allegations of discrimination and harassment need to be investigated and appropriate action taken. Employees must ensure that individuals who have or may have contagious diseases are not harassed or humiliated in the workplace. This protection would extend to individuals who have been quarantined because of SARS or who travelled to SARS hotspots. An employer must balance the obligation to ensure that the workplace is safe with the obligation not to discriminate or harass employees or allow such a climate to be fostered by other employees.

### *Dealing with a Human Rights Complaint*

For the most part, engaging in effective human rights practices is good for the proper conduct of business in your organization. In a public health emergency, employers will want to keep the lid on the situation. Employers should work hard to ensure that panic and a lack of perspective are not allowed to take root and distract everyone from the work that needs to carry on.

Once an employer becomes aware of possible discrimination or harassment there is a duty to investigate and determine whether a complaint is valid. If so, it must be remedied without delay. It is important to take detailed notes during the investigation and to retain the notes in case they are needed for a formal human rights complaint sometime in the future.

If the allegations of a complaint are not substantiated, the employee who brought the complaint should be advised accordingly, but should not be accused of any kind of dishonesty or bad faith in bringing the complaint, absent very clear evidence to that effect. Indeed, the employee should be encouraged to complain again if there is a similar incident in the future. A letter to the alleged harasser or the workforce in general also should be considered. The purpose of such a letter would be to remind employees of the company's discrimination and harassment policy, if one exists, and that any conduct contrary to the policy or the law would result in discipline. If such a policy does not exist, it is highly recommended that one be developed. The best time to develop a policy is when things are quiet, before there is actually a need for one.

When allegations of discrimination or harassment are substantiated, corrective action should then be taken. Dismissal is only an appropriate action in the most serious of cases. In a public health emergency, unless the employee who engaged in the misconduct was taking advantage of the situation or acting in bad faith, dismissal likely would be a harsh result. Such situations can be expected to breed a certain degree of fear, which can lead to poor judgments and comments or conduct that, later, will be regretted. Especially in these kinds of situations, an automatic penalty is not recommended, as each case has to be evaluated individually based on its own unique circumstances. Education and training certainly should be considered when there are allegations of discrimination and harassment in the workplace in the midst of a public health crisis. Generally speaking, training should be as well as, and not instead of, some disciplinary action, even if it is only a written warning that is issued.

### *Formal Complaint Process*

In the event that a human rights complaint is filed, the first step in most cases is for the parties to attend voluntary mediation to try and resolve the matter. If mediation is not successful an investigator will conduct an investigation and make a recommendation as to whether the complaint should be referred to the Human Rights Tribunal for adjudication.

If the Tribunal determines that discrimination or harassment has occurred it has very broad discretionary remedial powers.[7] The Tribunal can order an employer in breach of the Code to do anything required to achieve compliance. The Tribunal can issue cease and desist orders and payment for any losses suffered, including loss of earnings or loss of job opportunities. The Tribunal also has the power to order reinstatement, with or without back pay. Moreover, the Tribunal has authority to award a maximum $10,000 in damages for mental anguish where the infringement of the Code has been reckless or wilful. There is no monetary cap on the other types of damages that the Tribunal may order.

## Employment Standards

In accordance with the *Employment Standards Act, 2000*[8] ("ESA"), employees in Ontario who work for a company that employs at least 50 employees in Ontario are entitled to a maximum of 10 unpaid days off per year as emergency leave because of:[9]

- A personal illness, injury or medical emergency
- Death, illness, injury or medical emergency of a family member
- An urgent matter concerning a family member

These 10 days of emergency leave are inclusive of, and not in addition to, bereavement leave, sick leave or other family leave days made available through company policies and/or collective agreement provisions pertaining to leaves.

### *Definition of Family Members for Emergency Leave Purposes*

For purposes of emergency leave, a family member includes

- A spouse or same-sex partner;
- A parent, step-parent, foster parent, child, stepchild, foster child, grandparent, step-grandparent, grandchild or step-grandchild of the employee, the employee's spouse or the employee's same-sex partner;
- The spouse or same-sex partner of an employee's child;

7 *Ibid.*, s. 41.
8 S.O. 2000, c. 41.
9 *Ibid.*, s. 50.

- A brother or sister of the employee; and
- A relative of the employee who is dependant on the employee for care or assistance.

There is no statutory definition of relative; however, the *Employment Standards Act, 2000 Policy and Interpretation Manual*[10] provides that a relative is someone who is related through blood, marriage, same-sex partnership, spousal partnership, someone an employee lives with in a conjugal relationship, or through adoption. If the relative is not in one of the categories listed above, it must be established that the relative is dependant on the employee for care or assistance before there is entitlement to emergency leave.

There is also no statutory definition of "dependant", but generally speaking a relative will be considered dependant if he or she is reliant on the employee for some degree of care or assistance in meeting their basic living needs. Employment Standards policy is that a person does not need to be entirely dependant on the employee for care; there only has to be some degree of reliance. It is also not necessary that the person reside with the employee in order to be considered dependent.[11]

### *Communication Required for Emergency Leave*

An employee must inform the employer that he or she will be taking an emergency leave of absence. The ESA provides that if an employee has to begin an emergency leave before notifying the employer, the employee must inform the employer as soon as possible after the leave starts.

An employer is permitted to ask an employee to provide proof that he or she is eligible for an emergency leave of absence. The employee must provide evidence that is "reasonable in the circumstances." Evidence may take many forms including, without limitation, doctors' notes, death certificates, notes from a school or a day care facility or receipts.

Employees who take emergency leave days are afforded the same protections as employees who take pregnancy and parental leave under the ESA. The protections include the following:

- An employee on emergency leave continues to participate in any pension plan, life insurance plan, accidental death plan, extended health plan, dental plan, and any other benefit plan prescribed by

10 *Employment Standards Act, 2000 Policy and Interpretation Manual*, looseleaf (Toronto: Carswell) 18-10.

11 *Ibid.*, 18-11.

regulation, provided that the employee continues to pay his or her portion of the premiums, if any.

- Rather than forfeit vacation because he or she is on leave, an employee taking emergency leave may defer taking vacation until the leave expires, or until a later date agreed to by the employee and employer, or the employee may elect to waive entitlement to paid vacation time and simply receive vacation pay instead.
- When emergency leave ends, the employer must reinstate the employee to his or her former position, at the same rate of pay the employee would have been paid had the leave not been taken. If the employee's position no longer exists, the employer must reinstate the employee to a comparable position. However, there is no obligation to reinstate if the employee's employment was ended for reasons entirely unrelated to the leave. It should be noted that, in such a circumstance, the onus is on the employer to demonstrate that the reasons for the termination are completely unrelated to the leave. As an example, for employees in the hospitality industry in the spring of 2003, it is quite easy to conceive of a situation in which an employee in the midst of an emergency leave would have been selected, quite reasonably, as part of a group of employees for lay-off.
- An employer cannot intimidate, dismiss, or penalize an employee in any way, or threaten to do so, because the employee takes leave under the ESA, intends to take leave under the ESA, or will become eligible to take leave under the ESA.

Emergency leave is available to employees who have symptoms of West Nile, SARS or any other illness, but they are quite likely to be entitled to sick leave and perhaps even coverage under the *Workplace Safety Insurance Act*, as discussed below. More to the point, emergency leave is available to employees who need to stay home to care for designated family members. As well, if necessary, employees who are advised to enter home isolation or quarantine, and who are having difficulty dealing with their employer, may be able to cloak their absence under the emergency leave provisions. As long as the employee provides reasonable evidence to the employer that emergency leave is being taken for a legitimate reason, the employer cannot discipline or discharge an employee for the absence without being in clear contravention of the ESA. Most employment lawyers would relish the opportunity to represent an employee who is terminated from employment during a public health emergency simply because he or she was absent from work in compliance with quarantine requirements or to care for a sick relative.

If an employer were to discipline or discharge an employee for taking emergency leave the reprisal provisions of the ESA could apply and might be a better alternative than a wrongful dismissal action. If an Employment Standards Officer determines that an employer has violated the emergency leave and reprisal provisions of the ESA, any one of the following remedies may be ordered:[12]

- Compensation for any loss incurred as a result of the violation of the ESA
- Reinstatement
- Compensation for any loss and reinstatement

By providing for compensation and reinstatement as the best alternative for the offended employee, the reprisal provisions essentially give non-union employees the best protection in the event of dismissal that a unionized employee could expect. The reprisal provisions are also an effective avenue for a non-union employee to overturn an unjust suspension.

## Workers' Compensation

Workers' compensation in Ontario is governed by the *Workplace Safety and Insurance Act, 1997*[13] (the "WSIA"), which is administered by the Workplace Safety and Insurance Board ("WSIB"). To the extent that accidents occur in a workplace, or illnesses develop, there is a province-wide insurance scheme that provides benefits to workers on the one hand ("WSIB Benefits") and, on the other hand, protects employers from employee lawsuits.

### *Qualifying for WSIB Benefits*

To qualify for WSIB Benefits, there must be a worker-employer relationship that is covered by the WSIA. Most employees active in the province are covered. It also must be demonstrated that the worker has an injury or illness that is directly related to his or her work. There are a variety of WSIB Benefits, including lost earnings and medical aid.

12 *Supra* note 13, s. 104.
13 S.O. 1997, c. 16, Sch. A.

### *Employer Obligations Under the WSIA*

Workers' compensation has become focused on rehabilitation of injured workers and returning them to active employment, preferably at their former workplaces. An employer has legal obligations to accommodate such workers. Employers have a positive duty under section 41 of the WSIA to re-employ an employee who was in receipt of WSIB Benefits. This duty lasts until the earlier of the second anniversary of the injury or illness, one year after the employee is deemed to be able to perform the essential duties of his or her job, or until the employee reaches age 65. It also should be noted that if an employer terminates an employee within six months of returning to work after receiving WSIB Benefits, there is a presumption that the employer has not fulfilled its requirements under the WSIA. Therefore, it is incumbent on the employer to demonstrate that the termination was not related to the employee's injury.[14]

### *Occupational Diseases*

If an employee contracts West Nile, SARS or some kind of communicable disease, and if the employee is able to establish some reasonable basis for claiming that the onset of the disease was at the workplace, it is possible that the employee could claim WSIB Benefits. The WSIA provides compensation where, "a worker suffers from and is impaired by an occupational disease that occurs due to the nature of one or more employments in which the worker was engaged."[15] The definition of occupational disease is broad enough to encompass a disease such as West Nile or SARS, provided it can be demonstrated, on a balance of probabilities, that it was contracted in the workplace.

The WSIB is advising that workers with symptoms of SARS who believe that they were infected in the course of their employment may be entitled to WSIB Benefits. Each case will be assessed on the basis of its own individual facts. It is important to understand that you have to be sick in order to get benefits. WSIB Benefits will not be available for workers who are symptom free, even when quarantined or sent home on a precautionary basis. However, if a worker at home in quarantine or voluntary isolation develops symptoms of SARS, the worker may be entitled to

14 *Ibid.*, s. 41(10).
15 *Ibid.*, s. 15(1).

WSIB Benefits if the symptoms or an illness is compatible with his or her occupational exposure.

The WSIB also has advised that workers who become infected with West Nile virus in the course of their employment will be entitled to WSIB Benefits. Like SARS, and any other disease, entitlement will be determined on a case-by-case basis.

In the case of SARS, although we did not have statistical data at the time of publication, it can be fairly assumed that employers in the health care sector will be at risk of increased workers' compensation claims. Since there appears to be conclusive evidence of transmission of SARS in health care environments it is likely that health care workers who became ill with SARS symptoms during the outbreak will have WSIB claims approved. It will be more difficult for an employee in a non-health care setting to establish a claim for benefits under the WSIA respecting SARS. It is not enough to be a worker who had SARS symptoms; the disease must have arisen, on a reasonable analysis, from the worker's own workplace.

It seems obvious that employees who are required to work outdoors are at greater risk of contracting West Nile virus in the course of employment. However, even an employee who works indoors may be bitten by a mosquito carrying West Nile virus while at work or perhaps while sitting outside at lunch. Under either of those circumstances a claim for WSIB Benefits could be approved if there was evidence that the disease was a result of a mosquito bite in the workplace. It should be clearly understood that a lunch break taken at the workplace is a period of time that likely would be covered by the WSIA, so long as the worker is engaged in normal lunchtime activities. Accordingly, employers should take steps to prevent the risk of mosquitoes in the workplace, including the workplace grounds. Employers should be especially attentive to picnic areas at the workplace. Such precautions are discussed in more detail in the section about West Nile virus in Chapter 13.

## Privacy

Dealing with any kind of health issue in the workplace also requires an awareness of privacy issues. In Ontario, this continues to be a very unsettled area of the law. There is currently no statutory right to privacy in Ontario and the common law is somewhat murky. There is federal privacy legislation that will apply throughout Ontario by January 1, 2004, the *Per-*

*sonal Information Protection and Electronic Documents Act*[16] ("PIPEDA"); however, it will not apply to employment situations. There is still no provincial privacy legislation for employers in respect of their employees. Even so, the general advice for all Ontario employers is to not disclose medical information regarding its employees except in extraordinary situations and to keep any such information in a secure, private and confidential fashion.

SARS and other contagious diseases that are or could become serious public health issues are extraordinary situations. In such situations, the issue of what information an employer should disclose is often a difficult decision to make. On the one hand the employer needs to maintain the confidentiality of the employee who has potentially contracted SARS, but on the other hand the employer has to protect the health and well-being of other employees and members of the public who may have been in contact with a contagious employee. One of the issues that many employers grappled with was whether or not it should disclose the name of any employee who was self-isolated or quarantined as a suspected or probable SARS case. In the spring of 2003, the Ministry of Labour was advising employers to contact public health officials rather than disclose to the rest of the workforce, let alone the general public, the names of employees who were quarantined. This had no binding effect and was advice that was only as good as the ultimate advice from public health officials.

The federal privacy legislation provides that organizations cannot disclose personal information about employees without consent. Until January 1, 2004, it will apply only to personal health information for customers and employees of the federally regulated private sector and to cross-border disclosures of personal health information. After that date, PIPEDA will apply to Ontario and any other province that does not have comparable privacy legislation in place. Even so, its application will be limited to disclosures of personal health information to third parties and does not deal with use or disclosure of such personal information within the workplace. Although there have been a number of attempts to get privacy legislation off the ground in Ontario, it has yet to materialize. Presumably, this will be a matter dealt with by the incoming government following the provincial election in October 2003.[17]

---

16 S.C. 2000, c. 5.

17 The new Liberal government in Ontario has been unclear about its intentions in respect of provincial privacy legislation, at least to the point of its swearing-in, in late October 2003.

Whatever Ontario does will be built on the same 10 privacy principles that are the foundation for PIPEDA, so the federal legislation is relevant, both in the short-term, and as the likely model for the Ontario legislation. PIPEDA requires that an individual must know of, and consent to, the collection, use or disclosure of personal information by an organization in the course of commercial activities, except where inappropriate. This protection of personal information extends to the collection, use or disclosure of personal health information, subject to the following exceptions, which are very important to the subject matter.

PIPEDA has special provisions about the use and disclosure of private information that may apply in a situation where a highly contagious disease such as SARS may be present in the workplace. Given the Ontario experience, it has to be expected that any provincial legislation will follow suit. Section 7(2)(b) of PIPEDA provides that organizations may use personal information without the knowledge or consent of the individual(s) affected if there is an emergency that threatens the life, health or security of any individual. Section 7(3)(e) further provides that an organization may disclose personal information without consent to a person who needs the information because an emergency threatens the life, health or security of any individual. Section 7(3)(e) requires the organization to notify the person who the information is about in writing of the disclosure without delay.

Although there have been no rulings from the Federal Privacy Commissioner regarding section 7(2)(b) or 7(3)(e) of PIPEDA, arguably these sections would apply in the case of SARS or other highly contagious diseases in federal workplaces. In such situations, an employer may be able to rely on section 7 as a defence if it chose to disclose the name of an individual who had been in the workplace who was a probable or suspect SARS case.

Whatever might be the legislative situation now and in coming years, employers under both federal and Ontario jurisdictions should be cautioned that they should disclose personal information only to the extent that is reasonably necessary to protect the life, health or security of its other employees. Information regarding the health of an employee should remain confidential to the extent that is reasonably possible to do so. In the case of SARS or any contagious disease it is recommended that, rather than attempting to handle the situation in isolation, the employer should contact the local public health department or other public health officials and co-operate with such officials by providing information to them regarding persons who may have been exposed to the potentially ill and contagious employee. Public health officials should then take on the role of contacting anyone considered to be at risk. This way, employers are able to do their public duty, protect their workplaces as far as possible, and eliminate or at least limit liability in respect of issues arising from the disclosure of information.

Any medical information that is received by an employer does become company property. Access to employee medical information should be limited and on a need-to-know basis only. Medical information is one of the most highly sensitive types of personal information, and a company should utilize the highest security safeguards to maintain the confidentiality of medical information. Generally there is no reason why a supervisor should have access to employee medical information. A supervisor only needs to be advised of any required limitations, restrictions, or accommodation measures. Medical information should be kept separate and apart from personnel files. If at all possible, the company nurse or health services department should retain medical information and even should keep it confidential from other human resources employees.

# CHAPTER 4 — QUASI-CRIMINAL ISSUES

There are statutory provisions that come into play with SARS and other serious public health issues that can carry with them the threat of penalty for violation, including fines and even detention. These provisions are not actually criminal laws, but they are similar in character, which is why we describe them as quasi-criminal.

## Quarantine

In Ontario, the authority to place individuals in quarantine comes from the *Health Protection and Promotion Act*[1] ("HPPA"). Section 22 of the HPPA gives medical officers of health broad powers to restrict the mobility and conduct of persons. The HPPA gives a medical officer of health the power to order a person to take or refrain from taking a specific course of action in circumstances where there are reasonable and probable grounds that a communicable disease exists, or that there is an immediate risk that an outbreak of a communicable disease may exist. A medical officer of health's powers extend to ordering the following:

- Closure of certain premises, including a workplace
- Isolation of persons or groups of persons
- Cleaning, disinfecting or destruction of things
- Examination of a person by a physician
- Restricting the conduct of persons to ensure another person is not exposed to infection

The powers of the medical officer of health are not subject to the usual requirement of obtaining consent from the affected person.

A communicable disease is one that is prescribed by regulation made by the Minister of Health and Long-Term Care.[2] The list of communicable diseases is lengthy and ranges from food poisoning to influenza, to HIV/AIDS, Hepatitis, SARS and West Nile virus.

---

1 R.S.O. 1990, c. H.7.

2 See O.Reg. 558/91 under the *Health Protection and Promotion Act.*

## Obligation to Report

Under the HPPA, certain professionals have an obligation to report communicable or reportable diseases. Like a communicable disease, a reportable disease is also one that is designated by a Regulation made by the Minister of Health and Long-Term Care.[3] Both SARS and West Nile are considered communicable and reportable diseases pursuant to the Regulations under HPPA. Physicians, listed practitioners,[4] school principals, laboratory operators, hospital administrators, and superintendents of certain institutions are required to report communicable and/or reportable diseases to the medical officer of health.[5]

### Enforcement

Part IX of the HPPA deals with enforcement. It is under this section of the HPPA that an employee who breaches a quarantine order could be charged. Section 100(1) of the HPPA provides that any person who fails to obey an order under the HPPA is guilty of an offence. This section is broad enough to apply to individuals who fail to comply with quarantine orders as well as individuals and corporations who fail to report a communicable or reportable disease as required. Individuals who are found guilty of an offence may be fined a maximum $5,000 per day or part day that the offence continues. Where a corporation is convicted of an offence under the HPPA, directors, officers, employees or agents of the corporation who are responsible for the conduct of that part of the business also may be found guilty of an offence and may be subject to a fine of a maximum $25,000 for each day or part day that the offence continues.

The corporation, director, officer, employees or agents will be found guilty unless he or she can demonstrate that he or she took all reasonable care to prevent the commission of the offence. This is essentially a due diligence defence, just as required under the OHSA. As with prosecutions under the OHSA, prosecutions under the HPPA would be quasi-criminal in nature, and legal representation is recommended.

---

3 See O. Reg. 559/91 under the *Health Protection and Promotion Act.*

4 Chiropractors, Dentists, Nurses, Pharmacists, and Optometrists.

5 Sections 25-29 of the HPPA.

## Detention

Although the HPPA does not provide for jail time where the HPPA has been violated, there is a mechanism available for medical officers of health to apply to the Court to have individuals held and treated against their will if necessary. Where a person fails to comply with the orders of a medical officer of health, section 35 of the HPPA gives the medical officer of health the power to bring an application to a Judge of the Ontario Court of Justice. The Judge has the following remedies available:

- Order the person to be taken into custody and detained in hospital
- Order the person to be examined by a physician
- If determined to be infected, order the person to be forcibly treated

The initial period of detention is limited to four months, but another application may be made to the Court to extend the detention period if circumstances warrant further detention.

## Rules for Employers

The HPPA does not require employers to report that an employee has or may have a communicable disease, although in-house physicians likely will be required to comply with the reporting requirements of the HPPA. However, if an employer has knowledge or belief that an employee has a communicable disease that may put the public at risk, there is a moral obligation to inform public health officials who can then make the determination as to whether or not the matter should be investigated.

Where an employer is advised that an employee is under quarantine the employer should contact public health to ensure that there is no risk to other employees. Making inquiries regarding any procedures that should be followed is consistent with an employer's duty under the OHSA.

Under no circumstances does an employer have the authority to order an employee to quarantine or self-isolate. However, the employer can deny access to the workplace in appropriate circumstances.

◆

# CHAPTER 5 — STATUTORY ISSUES SPECIFIC TO SARS

In the event of a serious public health emergency, it can be expected that all levels of government will want to do something or, at least, will want to appear to be doing something about the problem, quickly and within the limits of their jurisdictional authority. This is what happened with SARS.

## SARS Emergency Leave

In response to the SARS emergency in Ontario, the provincial government, on May 5, 2003, enacted the *SARS Assistance and Recovery Strategy Act, 2003*[1] (the "SARS Act").

The SARS Act is retroactive to March 26, 2003, the day Premier Ernie Eves declared SARS an emergency in Ontario. The SARS Act provides for SARS emergency leave in addition to emergency leave under the ESA. The SARS Act does not specify the number of days of SARS emergency leave to which employees are entitled. Therefore, an employee is entitled to an indefinite leave of absence without pay, provided the requirements for taking the leave under the SARS Act are met.

SARS emergency leave provides affected employees with protections similar to the emergency leave provisions of the ESA. The SARS Act is intended to be a temporary measure, and the Lieutenant Governor in Counsel, at some unscheduled date in the future, will proclaim a day for which the eligibility for SARS leave will end.

The SARS emergency leave is in addition to emergency leave under the ESA. As the legislation is retroactive to March 26, 2003, any affected employee who took emergency leave under the ESA between March 26, 2003 and May 5, 2003 will be deemed to have taken SARS emergency leave instead of regular emergency leave, thus reinstating the employee's emergency leave bank.

Section 6(1) of the SARS Act sets out the circumstances under which an employee is entitled to SARS emergency leave:

- The employee is unable to work because he or she is under medical investigation, supervision or treatment relating SARS

---

1 S.O. 2003, c. 1.

- The employee is unable to work because he or she is acting in accordance with a SARS- related order under sections 22 or 35 of the *Health Protection and Promotion Act*[2]
- The employee is unable to work because he or she is in quarantine, isolation, or subject to a control measure in accordance with directions issued to the public, part of the public, or one or more individuals, by the Commissioner of Public Security, a public health official, a physician or a nurse or by Telehealth Ontario, the Government of Ontario, the Government of Canada, a municipal council or a board of health, whether through print, electronic, broadcast or other means
- The employee is unable to work because the employer directed him or her not to report to work in response to a concern the employee may expose others to SARS
- The employee is unable to work because he or she is needed to provide care or assistance to a family member as defined above under the Emergency Leave provisions of the ESA

Small employers must take note of the SARS Act. Unlike the emergency leave provisions under the ESA, SARS emergency leave is not limited to employers with 50 or more employees. This could set a very difficult precedent for small employers in the event of a SARS recurrence or other public health emergency. The SARS Act provides the same job protections, including reinstatement, as the ESA. Any employee subject to reprisals because of SARS emergency leave has access to the complaint procedure in the ESA.

An additional requirement of the SARS Act is that an employee who takes SARS leave must contact a physician or public health officials within two days to receive directions regarding whether or not the employee should continue to be absent from work. The employee is required to receive written confirmation of any such direction. An employee will not be entitled to SARS leave beyond the first two days unless a physician or public health official gives direction that the employee should not report to work. As with emergency leave, the employee is required to advise the employer that he or she is taking SARS leave as soon as possible. The employer is entitled to ask the employee for reasonable proof that he or she was entitled to SARS leave, but such proof need not be provided until after the leave has ended.

2 Section 22 and 35 of the HPPA deal with orders by Medical Officers of Health (MOH) relating to communicable diseases, and orders by the court when orders of the MOH are not complied with.

The ESA emergency leave provisions relating to continuation of seniority, benefits, vacation entitlement, reinstatement, and the prohibition against discipline and dismissal discussed above also apply to any leave taken under the SARS Act. However, neither the SARS Act nor the ESA restricts an employer's right to terminate or lay-off an employee if it is necessary to reduce the workforce because business has been adversely affected by SARS.

## Employment Insurance

The *Employment Insurance Act*,[3] which is federal legislation, provides for sickness benefits for a maximum of 15 weeks to qualifying individuals who are unable to work due to illness, injury or quarantine.[4] To be eligible for sickness benefits the employee must have worked 600 insurable hours within the last 52 weeks or since the start of the employee's last claim, whichever is shorter. Therefore, employees who are sick as a result of West Nile, SARS or any other illness, or an employee who has been quarantined can apply for Employment Insurance ("EI") benefits.

Normally, to qualify for sickness benefits an employee is required to provide a medical certificate.[5] Further, there is usually a two week waiting period before sick benefits are payable, unless the employee received paid sick leave from the employer during that period.[6] However, with SARS, many people elected to go into self-isolation upon the advice of public health officials, while employers sent others home. In many of these cases the employees were not actually ill and were not subject to a formal quarantine order.

In response to the uniqueness of the SARS emergency, the federal government amended the Regulation under the *Employment Insurance Act*, effective April 4, 2003, to loosen the eligibility requirements for sickness benefits in SARS-related cases. The revised rules provide that a medical certificate is not required where quarantine was:

(a) imposed upon an individual by a public health official, or

---

3 S.C. 1996, c. 23 as am.
4 *Ibid*., s. 12(3).
5 Can. Reg. 96-332, s. 40(1).
6 *Ibid*., s. 40(6).

(b) recommended by a public health official and the individual was asked by their employer, medical doctor, nurse or person in authority to place themselves in quarantine.[7]

The claimant is still required to provide a declaration that the period of quarantine was a result of either (a) or (b) above. The amendment also waives the two-week waiting period for sickness benefits where the quarantine is for reasons described in (a) or (b) above.[8] It should be noted that the amendment only applies to claims received and submitted on or after March 30, 2003. The new rules are only temporary measures intended to deal with the SARS emergency. Section 40(9) of the Regulation provides that the amendments to section 40 cease to have effect after six months. This means that the amendment relating to SARS only applies until October 4, 2003.

In addition to EI sick benefits, HRDC announced, in May 2003, a SARS Grant Initiative for health care workers affected by SARS. The additional funds are earmarked for health care workers who suffered a loss of employment income because they became ill or were required to be quarantined or isolated as a result of SARS, but were not eligible for EI benefits.

The special program provides weekly payments of $400 for full-time employees, and weekly payments of $200 for part-time employees. Eligible workers will be entitled to a maximum payment of $6000 for up to 15 weeks. The program is retroactive to March 30, 2003. Only employees who work in a recognized health care setting such as hospitals, medical offices and medical laboratories or emergency service workers such as paramedics, are eligible to participate in the program. The program is open to all employees who work in a recognized health care setting regardless of whether their job is medical related or not. The relief program is being delivered by HRDC's existing delivery network. Employees who believe they qualify for the relief should call 1-800-263-8364.

To be eligible for the special program, employees must meet the following criteria:

- The employee must be ineligible for EI benefits and demonstrate that the loss of income is not covered by some other form of compensation arising from employment, such as sick leave
- Employees must demonstrate a loss of employment income as a result of contracting SARS, being in isolation or under quarantine be-

7 *Ibid.*, s. 40(1.1).
8 *Ibid.*, s. 40(7).

cause they might have been exposed to SARS during work, or that they were prevented from working because of an outbreak of SARS in their workplace
- Any employee unable to work because he or she contracted SARS must provide a medical certificate confirming the employee had SARS and setting out the relevant time
- Any employee unable to work because he or she was in quarantine, in isolation, or prevented from working due to an outbreak of SARS in the workplace must provide a declaration confirming (a) the quarantine was imposed by a public health officer, or (b) the quarantine was recommend by a public health official for the general health and safety of the public, and that the employee was asked by his or her employer, doctor, nurse or person with similar authority to place themselves under quarantine
- Employees in isolation or prevented from working due to an outbreak of SARS in the workplace must also provide a declaration from the employer to that effect

The employee must also provide documentation from the employer stating that employment is in the health care setting and declaring whether or not the employee has full-time or part-time status.

## CHAPTER 6 — HUMAN RESOURCES ISSUES

Dealing with public health issues is not simply a matter of sorting out applicable legislation. There are common human resources issues that will affect all workplaces, whether unionized or non-union, and ranging from office settings to manufacturing plants or construction sites. Three significant issues that we want to cover are communication, dealing with absenteeism, and facilitating an effective return to work.

### Communication

One of the most important roles an employer can play when dealing with any public health issue in the workplace is that of a communicator. Every organization should demonstrate that it has strong leadership in times of emergency. One way to demonstrate leadership is by providing clear and frequent communications to the organization, in oral, written and, increasingly, electronic form. The communications must demonstrate that management has control of the emergency and it is taking appropriate steps to guide employees during times of uncertainty.

#### *Information*

Employees should be provided with clear direction regarding the public health situation, how it affects the workplace, and what to do. The information provided must be up-to-date, and follow the recommendations provided by public health officials. In addition to public health directions, employers need to be communicating any special health and safety precautions that are being introduced into the workplace. Employees also should be reminded of health and safety precautions already in place.

Providing instruction and training regarding the proper use of personal protective equipment or processes is a key part of the communication strategy. Employees also need to be advised of existing or new human resource policies that are applicable to the situation, such as attendance, compensation, and overtime policies. Employers should anticipate questions that employees will have and prepare question and answer type communications for distribution. This is often effectively done during town-hall meetings, either live or through an intra-net or private chat-room set up for that particular workplace.

### *Methods of Communication*

It is recommended that a consistent format for communication be developed that differentiates communications related to the medical emergency from any other company communication. The format can be as simple as using a different colour of paper or typesetting (font). It is also a good idea to keep communications relating to human resources issues separate from communications relating to public health advisories or health and safety precautions, and use a format that makes each type of communication easily identifiable to the employees.

The methods of communication will vary for each organization, and should also vary depending on the situation readily available. Organizations should consider the various methods of communication including bulletin boards, inter-office memorandums, intra-net postings, telephone hotlines, and face-to-face communication, and determine what is most appropriate for its workforce and the situation. These communication methods are not mutually exclusive. If employers face a SARS recurrence or another emergency similar to the SARS outbreak, a combination of the various methods of communication will likely be the most effective.

Another key aspect of communication is the ability to contact employees and visitors after they have left the premises. In the case of SARS, it was at times necessary to contact employees and visitors at home to advise of potential exposure to SARS, or to advise that the premises were under quarantine.

### *Contact List*

It is important that employers keep an up-to-date employee contact list that can be easily distributed to managers in an emergency so that employees can be contacted at home. Every organization also should be keeping track of who visits the workplace by having each visitor report to reception and sign in. Basic information such as the individual's name and telephone number should be collected for use in the event of an emergency. This sign in procedure also should apply to company employees who normally work at a different work site.

### *Dealing with Unions*

In a unionized environment it is very important to keep the lines of communication open with the union during an emergency. These should be situations that foster co-operation, not conflict. That said, in an emergency situation decisions often need to be made quickly, and some decisions may impact on the collective agreement. If the union is aware of the crisis and the impact it may have on the bargaining unit, there may be less resistance to the implementation of emergency measures, which may translate into fewer grievances during the emergency or after it is over.

## Managing Absenteeism

There are two main branches of the absenteeism problem. One branch is blameless or innocent absenteeism. Usually this means that the employee has no choice in the matter and is absent because he or she should not be at the workplace. Sometimes this simply means that management cannot prove that the employee was absent without good reason. Absences due to illness, including SARS and West Nile fall into the category of innocent absenteeism.

The other branch is culpable absenteeism, which features elements of misconduct, such as no-calls, no-shows, early departures and late arrivals. Culpable absenteeism generally involves absences that could have been avoided by the employee through the exercise of reasonable diligence and responsibility. Examples of culpable absenteeism include sleeping in and failure to arrange transportation to get to work. Casual absences that are unsupported by any medical verification may also be considered culpable where such absenteeism is unreasonable. Sometimes the two branches of absenteeism intertwine, such that the employee may be off for a valid reason, but has not called in or communicated with management in the appropriate fashion.

The first branch — blameless absenteeism — has two sub-branches. The first of these is inhabited by unfortunate employees who have long-term debilitating illnesses or injuries, including disabilities that may prevent them from ever returning to their previous position, or to any kind of work. The second sub-branch generally requires more action from management. This is where we find the employees who have repetitive short-term absences, often for a variety of alleged ailments, some of which may be claimed as arising from workplace situations.

Disciplinary action is not appropriate for any kind of blameless absenteeism, but there are non-disciplinary measures that management has at its disposal, and these often can be used to great effect to improve overall attendance, especially in respect of the second sub-branch. Culpable absenteeism, on the other hand, should be subject to the normal progressive discipline steps such as counselling, oral and written warnings, and suspensions.

### *Attendance Management*

Attendance problems become difficult for employers to manage when the employee has disability related absenteeism. A good absenteeism management program is needed to control both innocent and culpable absenteeism. The following practices will assist employers to determine whether or not absenteeism is excessive:

- Computer reports of the number of total days off and the number of separate instances of absenteeism within the last one month, three months and 12 months
- Identify the worst 10% (or other % below which there is a significant gap) in each time period
- Consider the reasons for the absence — one lengthy leave for medical reasons usually does not merit any employer response, even if the employee is in the worst 10%
- Deal with the worst cases first — the word will get around that "You are not going to get away with excuses for absenteeism forever"
- Progressive discipline does not apply to innocent absenteeism

### *Absenteeism Policy*

Absenteeism problems may be managed with the assistance of a good absenteeism policy. An absenteeism policy should contain the following:

- Statement of Attendance Expectations: The policy should describe the employer's philosophy regarding sick leave, absenteeism, and its commitment to abide by the *Human Rights Code*
- Reporting Procedure: The policy should set out the notification requirements such as the time frame for reporting, persons to contact, and whether the reason for the absence must be reported

- Medical Certificates: The policy should ask for medical certification from a qualified medical practitioner in certain circumstances, subject to the terms of the collective agreement, if any; for example, some provisions will require medical certificates only for leaves of more than three days, absent extraordinary circumstances
- Disciplinary Measures: The policy should inform employees that failure to abide by the policy without good reason could result in discipline, up to and including discharge

### *Human Rights Accommodation*

An employer has a duty under the *Human Rights Code* to accommodate an employee's disability to the point of undue hardship. In order to accommodate an employee, an employer has a right to be fully informed of an employee's work limitations and any modifications required to the employee's job duties. In order to properly manage absenteeism it is imperative that the employer should take consistent action to enforce its right to receive proper information and notification from the employee.

### *Information Relevant to Accommodation*

An employee has a duty to co-operate in the accommodation process, as does a union representative if there is one. An employee does not have to disclose his or her medical diagnosis to an employer, but the following information must be provided to an employer:

- Details of any functional limitations experienced by the employee that pose a conflict with workplace requirements
- The employee should be providing information regarding the nature and degree of any physical restrictions and the duration of any restrictions
- Details of how any treatment needs, medication or medical interventions affect participation in the workplace, such as the scheduling of medical appointments where conflict cannot be avoided with regular work hours
- In the context of total disability, details of the prognosis in terms of timing for return to work or at least some indication of when further information in that respect might be available
- In long-term absences, the employee should be contacting the employer every time that there is a material change in treat-

ment, recovery, or projection of a date for returning to full or modified employment

While there is some obligation on employees to maintain good communication, employers should take active steps to obtain the required information, especially in cases of repetitive or long-term absenteeism. This means that the employer should advise the employee and the medical practitioner in writing as to what is required. The letter should confirm that the employer is writing to obtain information regarding the employee's accommodation issues. The letter should also advise that it is the employee's responsibility to facilitate the employer's access to information regarding accommodation.

The employer should provide the medical practitioner with a questionnaire to be completed. A copy of the employee's job description or a physical demands analysis should also be provided to assist the medical practitioner to complete the questionnaire. The questionnaire should seek information regarding the following:

- Details of functional limitations
- Scope of conflict between functional limitations and the employee's job duties
- Projected timeline for persistence of functional limitations
- Identification of any treatment needs or medication that will affect, in any way, the work responsibilities of the employee
- The prognosis

### *Problems with Co-Operation*

If the medical practitioner and/or employee fail to co-operate with the employer, the employer should write to the parties advising that the employee has an obligation to provide the requested information. The letter should also advised that failure to comply with the request may risk the employee's right to accommodation and that the employee may be disciplined for persistently refusing to provide required information.

That said, in the wake of the recent Ontario Court of Appeal decision of *Prinzo v. Baycrest Centre For Geriatric Care,*[1] employers need to ensure that all communications with employees absent due to disability are respectful and do not constitute harassment. In that case, Prinzo was awarded punitive damages because it was determined that she suffered

1 (2002), 60 O.R. (3d) 474 (Ont. C.A.).

anxiety and emotional upset as a result of her employer persistently contacting her in regard to her work-related injury.

Prinzo was 49 years old and had been employed with Baycrest for 17 years. Shortly before receiving a layoff notice, Prinzo had suffered injuries in a slip and fall incident in the employer's parking lot. Prinzo was absent from work for several months and was unfit to perform any work during this time. While she was absent, her employer persistently urged her to return to work, at one point falsely implying that her doctor had agreed that she was fit to return to work. Prinzo's lawyer wrote to the employer describing the stress and anxiety its conduct was causing Prinzo, and requested that all future communication be directed to him. The employer continued to contact Prinzo directly. When Prinzo did return to work the employer met with her alone for two hours and insisted on talking about termination despite Prinzo's protest that she was in no condition to address the subject. Prinzo was eventually given notice of termination. The trial judge awarded Prinzo damages for harassment and 18 months' pay in lieu of notice. The Ontario Court of Appeal reduced the notice period to 12 months, since Prinzo knew well in advance that her employment would end. However, the Court of Appeal upheld the award of $15,000 for the intentional infliction of mental suffering.

### *Proper Documentation*

All initiatives to manage access to the information required and the employer's responses to an employee's non-compliance with requests should be well documented. Any documentation produced by the employer should be copied to the employee if the employee is not already the recipient. Without proper documentation it will be difficult to justify a contention that an employer's accommodation obligation has been frustrated by the employee's conduct.

### *Dismissal for Blameless Absenteeism*

Although an employee cannot be disciplined for blameless absenteeism, there does come a point when it may be appropriate to dismiss an employee as a result of persistent blameless absenteeism. If done correctly, such dismissals stand a reasonable chance of succeeding without giving rise to a legitimate claim for damages or human rights complaint. As with many employment situations, but especially because these are dismissals that have human rights and possibly other statutory implications, the employer probably should get legal advice from the outset, or at least before taking the final action.

### *Indefinite Absenteeism*

If an employee is ill or injured and absent from the workplace for an indefinite period, there comes a time when the employer can fairly conclude that there is no reasonable expectation that the employer will be able to accommodate medical restrictions. It is typical to wait two years before making such a determination. The period of time may vary according to disability plans and workers' compensation situations, whichever is applicable.

For most cases, there is a two-step dismissal process:

1. Make sure that you get current medical information. Let the employee know, by letter, what medical information you have on file, and that the information indicates that the employee is unable to return to work. Invite the employee to provide further or different medical information. Make clear to the employee that, unless there is information that indicates the realistic possibility of a return in the foreseeable future, then employment will be terminated. Clarify the status of any disability coverage. Provide the employee with a timeframe for response, typically two weeks, and indicate that, if there is no response, you will assume that your information is correct.
2. Act promptly on whatever response you get, including a failure to respond within the timeframe. If the information suggests that employment may resume in the near future, then you should become very active on the file, to ensure that you are not simply being placed into an indefinite holding pattern. Insist on dates for further medical reports, within weeks, not months, and consider getting an independent medical review or even an independent medical examination if you have concerns about the quality of the medical reporting.

### *Recurring Absenteeism*

The other kind of non-performance due to absenteeism is when an employee is frequently absent for short periods. Dealing with this kind of absenteeism becomes complicated if some of the absences are due to workplace injuries or illnesses. As well, Ontario's emergency leave provisions, discussed above, effectively provide employees with a free ride for a maximum of 10 days per year. Further, for SARS-specific absences an employee is entitled to the SARS emergency leave which protects an employee's job. That said, for most cases, there is a three-step dismissal process, with each step featuring a meeting and a follow-up letter:

1. Introduce the problem, with statistics going back a year or two. You will need good attendance records, and you should relate the employee to his or her own group (such as a production unit or an office unit). Indicate that the employee's attendance is below the norm. Clarify that this is not disciplinary — you are not challenging the reasons for absence. Recommend that the employee get medical assistance to deal with any underlying problems. Provide the employee with a timeframe for improvement, typically two or three months. Warn that failure to improve attendance could lead to dismissal. Follow up with the employee at the end of that period or in the meantime, if the problem continues to be serious.
2. Follow up. If the problem is the same or worse after the first monitoring period, then you should provide a final monitoring period of the same length (two or three months). Make it as clear as possible that a failure to improve attendance will likely result in dismissal.
3. Dismiss or, at least, follow up for one final time at the end of the second monitoring period. In some cases, you might give one last chance to an employee. A proper exercise of managerial discretion requires that each case be considered on its own merits. If an employee appears to be making a real effort to improve attendance, then you should be patient. As with progressive discipline, when you are dealing with a situation of frequent short-term absenteeism, the primary goal is to improve the employee's attendance, and not to terminate employment unless absolutely necessary.

Employers should be cautioned that, before terminating an employee because of frustration of contract related to chronic absenteeism, the employer must ensure that it has explored all accommodation issues prior to making such a decision. There have been some disturbing indications from some Canadian adjudicators that there will be an extreme tolerance for absenteeism, depending on the particular facts.

For example, in two recent decisions,[2] the Canadian Human Rights Tribunal (the "Tribunal") took the position that accommodation may require tolerance of excessive absenteeism, at least where the employer's operation is large, and the skills of the employee are not unique. These two *O-C*

2 *Parisien v. Ottawa-Carleton Regional Transit Commission*, (2002) 44 C.H.R.R. D/94 (Can. Human Rights Trib.); *Desormeaux v. Ottawa-Carleton Regional Transit Commission*, [2003] C.H.R.T. No. 1, 2003 CHRT 2 (Can. Human Rights Trib.) [hereinafter referred to as the O-C Transpo decisions].

*Transpo* decisions both involved bus drivers. Ms. Desormeaux was absent on average 6.5 days per year over a period of 8.75 years due to migraines. Mr. Parisien was absent due to illness more than 1600 days over a period of 18 years. Both employees were discharged for excessive absenteeism. In both cases the Tribunal determined that it would not be undue hardship for the employer to accommodate the intermittent absenteeism, particularly since O-C Transpo employed a large workforce, and the bus driver duties were interchangeable.

There was some encouragement for employers, as Tribunal Member Ann Mactavish stated that she could accept that in some situations, "intermittent absenteeism could potentially create undue hardship for an employer where, for example, a small workplace was involved, and the individual in question provided unique services." Even so, in these cases, the Tribunal concluded that the employer had not sufficiently explored alternative jobs within the organization or other accommodation options. The Tribunal reinstated both employees.

### *Absences Due to Quarantine*

Absences related to SARS proved to be quite challenging for employers in the spring of 2003. Employers had to deal with employees absent because they had been ordered into quarantine pursuant to the *Health Promotion and Protection Act*, employees absent because they met the criteria for self-quarantine, and employees absent on their own election because they were afraid of contracting SARS if they reported to work.

Employees ordered into quarantine by medical health officials clearly had a valid reason for being absent from work, and the employees affected were legitimately able to take emergency leave under the *Employment Standards Act, 2000*, or SARS emergency leave under the *SARS Assistance and Recovery Strategy Act*.

Human rights legislation arguably applies to employees quarantined as a result of SARS even if they do not experience symptoms of SARS. Therefore, an employer may have a duty to accommodate quarantined employees by permitting them to work from home if at all possible or at least by accommodating their absence from work.

It is relatively easy to substantiate absences as a result of a quarantine order under the HPPA. An employer can simply ask the employee for a copy of the written order from the medical officer. As discussed above, written orders by a medical officer can be enforced by the Courts and result in significant penalties if not complied with.

In the case of SARS, public health officials recommended self-quarantine to individuals who meet established criteria. The criteria changed throughout the SARS emergency and had to be monitored on a daily basis in order to determine who should self-quarantine. Those employees who legitimately meet the criteria for self-quarantine should be entitled to the same protections as the employees who were ordered into quarantine pursuant to the HPPA. However, substantiating the legitimacy of a self-quarantine may in some cases be more difficult.

Employees who choose to self-quarantine will have evidentiary problems as to the reasonableness of their actions by comparison to employees who are ordered into quarantine. Employers should attempt to validate the employee's decision to self-quarantine; however, to a certain extent, the employer may have to take the employee's word that he or she does meet the criteria for self-quarantine. A doctor's note could substantiate the absence, but in a situation where the employee does not exhibit any symptoms of SARS a doctor's note may not be of any value. An employer could ask questions of the employee regarding his or her attendance at the locations where the employee was possibly exposed to SARS, but it may not be possible to obtain objective proof that the employee actually meets the criteria.

Another group of employees who are a challenge to deal with are those who do not meet the criteria for self-quarantine, but choose not to attend work because of fear of contracting SARS. With SARS the information that was provided by health officials indicated that there was no real risk to most employees. In such a situation employers are entitled to expect employees to report to work unless they meet the criteria for self-quarantine or are subject to a quarantine order. Fear of SARS or any other contagious disease alone likely would not be sufficient to justify absence from work. Even so, the work refusal procedures under the OHSA must be complied with before disciplining or threatening to discipline employees who refuse to attend work because they feel it is unsafe. There is a fine line to walk in a public health emergency between sympathy for the natural fears of workers and the need to insist on a proper respect for the needs of the organization.

### *Consequences of Attending Work While Sick*

The SARS emergency has highlighted the problems associated with employer policies aimed at minimizing employee absenteeism. Although an employer has a legitimate right to expect employees to report to work, the organization does not benefit when employees report to work sick. Business pressures or pressure to avoid financial loss or penalty motivates many employees to report to work even if they have a bad cold, the flu, or other contagious diseases.

The result of employees reporting to work when not feeling well is demonstrated by what happened in the health care sector. Most of the employees affected by SARS were workers in the health care field. Patients infected some workers, but in the early stages of the SARS outbreak, many were infected by colleagues who reported to work even though they were feeling sick.

A further example of the consequences of a sick employee reporting to work is the situation experienced by Hewlett-Packard in Markham, Ontario. One Hewlett-Packard employee who was supposed to be in voluntary quarantine chose to report to work. Public health officials had not advised Hewlett-Packard that one of its employees was in voluntary quarantine. As a result of one employee reporting to work when he should have been in quarantine, the entire building was ordered closed by public health officials. Approximately 200 Hewlett-Packard employees and visitors to the building were put in 10-day isolation, which was a significant setback to the business.

Employers should consider advising employees that employees who break quarantine to attend work, or who report to work when they know or ought reasonably to know that they have been exposed to SARS, will be subject to discipline up to and including discharge.

It also should be communicated to employees that reporting to work when they have been ordered into quarantine may be an individual violation of the *Occupational Health and Safety Act*, and such conduct may also result in quasi-criminal proceedings.

SARS has highlighted the serious consequences of employees reporting to work sick. However, there are many other contagious illnesses that may not have life or death consequences, but are still spread from worker to worker. Every employer should consider taking a look at how it manages its sick leave policy. What message is being sent to employees? If that message is that employees should be coming into work even if they have colds or are suffering from the flu, then perhaps the message needs to be changed. The costs of increasing the number of sick employees is likely greater than the cost of having one or two employees stay home for a day or two to recuperate.

An appropriate sick plan, both short-term and long-term, may be a prudent investment. In the exceptional circumstance of a public health crisis, the employer should consider making at least a partial commitment to maintaining wages for employees in particular situations, like quarantine. We recognize that it is a tough balance, as employers do no want to encourage unwarranted absenteeism. With this balance in mind, practical encouragement to employees to comply with sensible health precautions would go a long way to protecting the rest of your workforce. The choice that is made in the public good, though it may have a price tag, may be a much less expensive choice than an undue pressure on employees to attend work without proper regard for the risk of contamination

## Dealing with Return to Work Issues

It is important to note that an employer does not have access as of right to an employee's private medical records. Absent collective agreement or contractual language, an employer has no right to require a medical diagnosis. The only information that an employer has the right to know is whether or not an employee is fit to perform the duties of his or her job, and whether or not any accommodation is required.

Further, absent a statutory authority or collective agreement or contractual language, an employer cannot compel an employee to be examined by a physician who is not the employee's own physician. The statutory exceptions to this rule include the situation where the employee is in receipt of benefits under the *Workplace Safety Insurance Act*, or where an employer is required to set up a medical surveillance program under the *Occupational Health and Safety Act*. Employees may voluntarily agree to independent medical examinations, either to co-operate with the employer or because the continuation of sick benefits may depend on it. Independent medical examinations may be a reasonable requirement if there is doubt about the medical situation, and may be an alternative for an employee to threats of dismissal or to a continuing absence when the employee feels ready to return.

When an employee has been off work due to illness, an employer can presume that the employee is unfit to perform his or her duties. The employee's absence may continue past the point when the employee feels ready to return if the employer is not satisfied with the medical evidence of fitness. Employers do have the right and obligation to determine whether or not an employee is fit to return to work. In the case of an employee who has been absent due to SARS or another contagious disease, the duty will extend to obtaining assurances that the employee is not a danger to others through his or her condition.

An employer is entitled to require a medical certificate from the employee's doctor regarding the employee's fitness to perform the job. The onus is on the employee to establish that he or she is fit to return to work. If an employer has reasonable grounds, it may reject a medical certificate and ask for additional information, such as an independent examination. Circumstances where a medical certificate reasonably may be rejected include situations where an employee had a lengthy medical absence and provides a one-line prescription pad note from his or her doctor. If the medical certificate is ambiguous or equivocal, or contradicts a recent medical certificate, an employer should consider requesting clarification of the medical certificate. Again, these are the kinds of situations that are ripe for the employer to request an independent examination, especially if the response is poor to the request for additional information.

Where an employer does reject a medical certificate provided by an employee, the employer must state the grounds for the rejection. The employee should be told what is required before he or she is permitted to return to work. It is also advisable to provide a job description to the physician so that appropriate accommodation measures may be recommended.

Employers should be persistent. Make sure that you get the information that you require before absent employees are allowed to return. It is employers who will shoulder the blame if problems develop following a premature return, whether with the returning employee or other employees in the workplace.

◆

# CHAPTER 7 — COMPENSATION ISSUES

Although employers are not required to compensate employees unable to work due to illness or quarantine, making a commitment to maintain wages during a public health crisis goes a long way towards encouraging employees to respond in a safe manner.

### *Short-Term Disability Benefits*

An employee who is ill as a result of SARS or West Nile likely will qualify for short-term disability benefits, provided the employee provides the medical documentation required to substantiate the illness. In the case of contagious diseases such as SARS, an employee may not be ill but still be required to remain in quarantine as a precaution. Most short-term disability plans do not provide for coverage under such situations. An employer should contact its insurer to determine whether the policy covers quarantine situations where the employee may not be ill. In some cases the insurer may be willing to cover claims on an extra-contractual basis provided the employee meet the conditions for self-quarantine set by public health officials.

For employers that are self-funded for short-term disability, the terms of the policy should be reviewed to determine whether the policy would extend coverage to employees who are quarantined, but not ill.

### *Paid Leave of Absence or Use of Vacation*

Where there is no short-term disability program available to employees, an employer can also consider a paid leave of absence. This is by no means a legal obligation, but it may be a good human resource practice if it encourages possibly contagious employees to remain at home. Where a paid leave is not practical, an employer in Ontario does have the right to require the employee to take vacation days provided that it is in blocks of at least a week at a time.[1] This option prevents the employee from attending the workplace, but does not deprive the employee of income for the period of quarantine.

---

[1] *Employment Standards Act*, 2000, S.O. 2000, c. 42, s. 34.

### *Working from Home*

Depending on the nature of the employee's position it may be possible to work from home during quarantine. Where an employee is healthy, but unable to leave home for an extended period of time, the opportunity to be productive likely will be welcomed. Working from home also will allow the employee to continue to be paid for the duration of the quarantine. The option of working at home will be discussed in more detail in Chapter 9.

### *Leave of Absence Without Pay*

Where none of the options above are available, an employee should be granted a leave of absence without pay for the period of quarantine. As we have discussed, the emergency leave provisions of the *Employment Standards Act, 2000* apply not only to employees who are in quarantine, but to employees who are caring for dependents who are ill or quarantined due to SARS. As well, in the case of SARS-related absences, the SARS emergency leave is also applicable until the Lieutenant Governor in Counsel proclaims a day for which the eligibility for SARS leave ends.

### *Employment Insurance Benefits*

Employees should be advised that they could apply for Employment Insurance Benefits for the period of quarantine. As previously discussed, employees who are required to be in quarantine are entitled to a maximum of 15 weeks of special benefits if unable to work because of illness, injury or quarantine. Chapter 5 outlines the details of the changes to Employment Insurance benefits specifically for individuals in SARS-related quarantine.

### *Compensation Issues for the Quarantined Workplace*

It is possible that an entire workplace could be subject to a quarantine order. This is most likely to occur in workplaces such as schools or hospitals. This poses an interesting problem for employers who have employees on site 24 hours a day, 7 days a week. In such a situation, an employee is entitled to pay only for hours actually worked during the time of the quarantine. This is subject to any collective agreement that may be in place, although it would be unlikely that the situation would be caught by such an agreement. Although the hours of work and overtime provision under the *Employment Standards Act, 2000* would still apply in the circumstances where employees were unable to leave the workplace due to quarantine, employees are not entitled to overtime pay just because they are unable to leave the workplace. You have to work to be paid.

Simply having to remain at the workplace does not meant that you are on duty and at work. An employer should be careful, in such a situation, to have a clear differentiation between on-duty and off-duty employees. In any event, an employer should do its best to provide for off-duty employees who are compelled to remain at the workplace due to quarantine.

◆

# CHAPTER 8 — CONTINGENCY PLANNING FOR EMERGENCIES

After September 11, 2001, many employers who did not have a business continuity plan in place realized the importance of preparing for previously unimaginable disasters, and took steps to implement one. Although a small company will not have the same resources to dedicate to emergency planning as a large company, a minimal plan can be put in place without incurring overwhelming costs. The question that needs to be asked is whether or not the company can afford not to have a contingency plan in place. The cost of an extended business interruption can be much more devastating than the cost of implementing a contingency plan.

Living through the SARS outbreak has emphasized that contingency plans are not only needed for man-made and natural disasters, but also for health-related emergencies. A business interruption as a result of a health-related emergency could result in loss of business to competitors, as well as a loss of reputation for the business, human resource issues, and health and safety liabilities. Any contingency plan that is developed needs to comply with applicable laws, protect employees and the community, and ensure that the organization has the ability to recover from a business interruption.

The management team that is comprised of the key leaders in the event the contingency plan is utilized should be the team to develop the contingency plan. It is critical during any type of emergency situation that management knows what to do so that the other employees know what to do, including who to go to for guidance and direction. Any contingency plan must specify who is responsible for making decisions. The decision makers must then be trained in order to be familiar with the contingency plan and employees must be advised as to who are the decision makers.

The Human Resources department should play a key role in developing the contingency plan. Human Resources is a source of information about the employees and is well equipped to give advice regarding the practical and legal implications the plan may have on the human resources of the organization.

In the event of a health-related emergency, it is possible that many staff will be unable to report to work due to illness or quarantine. It is therefore important that any contingency plan consider how the business will continue to operate without a full compliment of staff.

At the very least, there needs to be a communication pyramid set up, so that employees can be quickly contacted in the event of an emergency and throughout its duration.

Another recommendation is to develop a staff skills profile so it is easy to identify who can fill in for an employee absent due to illness or quarantine. This may not be critical when only one or two employees are missing, but when you are dealing with large numbers of absences it will be necessary to know who can cover for whom. Although everyone likes to believe their job is critical, the reality is that in a crisis situation a business can survive without some non-essential duties being performed.

Every organization has certain jobs that are critical to the organization and that must be performed in order to stay in business. The key positions in each department should be identified and back up plans implemented in the event that key individuals are going to be absent for prolonged periods of time.

The management team responsible for the contingency plan should consider every worst-case scenario related to business continuity and then should develop a plan to deal with each situation. The better prepared the business is, the less likely an interruption in business will occur. At a minimum, the organization will be able to respond to business interruptions quickly and avoid lengthy interruptions. A proactive approach is usually better than a reactive approach.

SARS was the first widespread medical emergency that modern employers in Ontario had to deal with in the workplace. Although almost all organizations seemed to make it through the crisis, the reality is that many, if not most, employers were caught off guard and were addressing issues as they arose. Employers in the health care sector faced the brunt of the SARS emergency. If the next health care emergency is not as isolated as the SARS emergency the reality is that the impact on many more businesses could create a true economic crisis. In such a situation, the recovery for individual businesses and organization, as well as for the entire local economy, could be long and painful.

Employers should take the opportunity to be prepared for the next health-related emergency. If the organization has determined in advance how to deal with certain situations, its response to the emergency will be quicker and more effective. Further, more time can be spent on the new issues that arise with the next health-related emergency rather than spending time making decisions on issues that could have been dealt with in a contingency plan.

When making a contingency plan it is critical that the long-term impact of the decisions be taken into consideration. In a unionized environment the impact of decisions on the collective agreement need to be considered. Look to the collective agreement for guidance with respect to how much leeway you have to change policies and procedures in the event of an emergency. Consider keeping the union in the loop when decisions are being made that affect the collective agreement, or be prepared to deal with the grievances after the fact. There well may be collective bargaining issues that should be addressed during negotiations.

Some of the issues that should be considered and included in a contingency plan related to health emergencies include:

- *Alternative arrangements if the building is quarantined* — If the building is quarantined, employees will not be permitted to report to work or in some cases employees will not be permitted to leave the building. Is it possible to continue operating the business from an alternative location? Can staff work from home? Can the business continue to operate with staff that is quarantined in the building?
- *Leave policies* — Ensure you understand the law with respect to absences from work. Decide in advance what will be considered sick time and under what circumstances employees will be granted paid time off.
- *Human Resource policies* — Identify what human resource policies may be affected and need to be modified in the event of an emergency, such as overtime policies, scheduling issues, payment for cancelled shifts, payment of employees who are not permitted to leave the building. Consider the need to revise travel policies or establish evacuation procedures for staff deployed abroad.
- *Organizing remaining staff* — It may be necessary to redeploy staff from other departments or locations. Know what skills your employees have and what positions are critical.
- *Replacement workers* — Identify what staff is essential and will need to be replaced in order to continue operating. Identify how you are going to replace a large number of staff in a very short period of time and how you are going to train them.
- *Working from home* — Identify who can work from home and assist those employees to set up in order to be able to work from home.
- *Communications* — Identify who will communicate information to the public, customers, and staff. Determine what type of information will need to be communicated and the best method of communica-

tion. Ensure there is a system in place to call staff, and ensure managers have current employee contact information readily available.

- *Health and Safety* — Ensure that training is available to staff where new Personal Protective Equipment is needed. Ensure all employees receive training regarding their respective duties under the *Occupational Health and Safety Act*.
- *Counseling* — Provide information to employees regarding counseling services available during the emergency. Communicate information regarding existing employee assistance programs ("EAP") and consider the availability of additional counseling services in the workplace for the duration of the emergency.

As we have stressed, public health officials have warned us that SARS is not yet eradicated and may rear its ugly head again in the future. There is also concern over other unknown diseases that may make an appearance in North America. In order to ensure that your business survives the next health care emergency, it is essential that a contingency plan be implemented sooner rather than later.

◆

# CHAPTER 9 — WORKING FROM HOME

## Benefits of Working from Home

Today it is very common for employees to do at least some of their regular work from home instead of or in addition to going into the office. Advancements in technology have made it possible for employees to be connected to the office via computer, telephone, fax, and e-mail. This type of arrangement is called telework or telecommuting. Some companies have formal telecommuting policies and procedures in place; others have *ad hoc* arrangements, whereby some employees are able to access computer systems and e-mails from home, often after hours and on weekends. There are many benefits to telecommuting in the context of a public health emergency, including a flexible work arrangement that may assist the quarantined or disabled employee to remain productive when unable to physically attend the workplace.

Some form of telecommuting is a concept that should be considered by employers because of the benefits it provides to both employees and the company. For employees, working from home may provide higher job satisfaction, fewer distractions and better personal time management; it may save time and will save commuting costs. For the company, benefits may include higher employee retention, higher productivity, reduced absenteeism, and savings in office space requirements. In a public health emergency, telework provides a solid backup system that may allow the business to continue functioning at a much higher level of effectiveness than its competitors.

Telecommuting is certainly not without disadvantages. If telecommuting is not implemented properly there may be feelings of isolation by home workers, a lack of separation of family and work, possibly more distractions depending on the household, possible delay in service to customers, and difficulty communicating with employees. In public health emergencies, poorly thought out procedures may simply add to the chaos.

It is important to note that many positions and employees will not be suitable candidates for telecommuting. An employee who works on the assembly line or who works in a hotel or restaurant or in a customer service capacity is not going to be able to telecommute, unless he or she is assigned different duties. However, many office workers ranging from the senior executive to a data entry clerk might be prime candidates to telecommute on a regular basis or as a temporary measure to accommodate a quarantine or disability.

### *Planning Ahead*

Since it is impossible to predict when an employee will need accommodation as a result of quarantine or disability, a prudent employer will review the possibility of telecommuting at its workplace and identify what needs to be done in order to effectively implement telecommuting. If the infrastructure for telecommuting is put in place now, the transition to telecommuting will be seamless when it is required in the event of a health-related emergency.

In a true telecommuting situation an employee would perform all of his or her work responsibilities from a workstation at home, and would attend at the workplace on an as needed basis to attend meetings. However, there are many variations of telecommuting including situations where employees only work from home on specific days, or on an as needed basis only.

### *Duty to Accommodate*

In a situation where an employee is disabled or under quarantine, the duty to accommodate under the *Human Rights Code* may require the employer to permit the employee to work from home despite the fact that the employee cannot perform every job task at home. Provided the essential duties of the job can be performed from home, telecommuting likely would be required unless the employer can demonstrate that it would create an undue hardship.

Employers can evaluate requests to permit telecommuting on a case by case basis. An employer who reviews the types of jobs within the organization with a view to identifying what types of jobs can be done at home will be better prepared for the next emergency situation such as SARS. Permitting telecommuting provides the employee with income protection while the business continues to operate with minimal disruptions.

## Working from Home Policies

Generally it is good practice to regard telecommuting as any other term of the employment relationship. It is therefore a good idea to set out clearly in writing a telecommuting policy. While the language of such a policy will vary from employer to employer, it is recommended that the policy should include the following:

(a) Define what telecommuting is and clearly set out duties, expectations and deadlines
(b) Make it clear that telecommuters are still employees, and are therefore subject to the employer's policies and procedures
(c) Set out the work hours and days for telecommuters
(d) Advise that the employer may modify or cancel the telecommuting arrangement at any time without cause or notice
(e) Notify the telecommuter of his or her responsibility for maintaining a safe work environment
(f) Grant the employer the right to carry out periodic inspections of the telecommuter's work area, and state when and by whom such inspections will take place
(g) Notify the employee that workers' compensation insurance applies to telecommuters and require that all work-related injuries be immediately reported
(h) Explain where the employee may hold business meetings
(i) Assign liability for injuries to third parties which occur in the telecommuting workplace
(j) List in detail what equipment the telecommuter will use, who will provide and maintain it, where the equipment will be located, who will be responsible for loss or damage and for insuring it, and when any employer-owned equipment must be returned if the employment relationship ends
(k) Notify the telecommuter of his or her duty of confidentiality, setting out any procedures to be followed to ensure that the duty is not breached

The above list is illustrative and not intended to be exhaustive; it will depend on the particular needs of the employer. Furthermore, in situations where telecommuting is a form of accommodation, the policy may need to be modified. For example, establishing strict hours of work may not work in a situation where the employee who is working from home is disabled. One of the advantages of working from home for a disabled employee is the ability to work when the employee is feeling well enough. Provided the work is being completed an employer may have to demonstrate some flexibility in that regard.

### *Supervision*

One of the factors that should also be considered before implementing telecommuting is the issue of supervision. When dealing with long-term or full-time telecommuting situations, arrangements should be made to ensure adequate supervision of the telecommuter. This includes determining the frequency of contact between the employee and supervisor, on-going feedback, and performance reviews.

### *Telecommuting Agreement*

Where an employee is permitted to telecommute on a full-time or long-term basis it is recommended that the employer and employee enter into a written telecommuting agreement that includes the provisions discussed in the section of working at home policies, above.

In most cases, a telecommuting arrangement should be offered on a trial basis for a specified period of time. The policy should clearly state what criteria would be used to evaluate the arrangement. Evaluation may include the following items:

(i) meeting deadlines;

(ii) employee productivity;

(iii) progress of individual or team assignments;

(iv) availability to receive and return calls;

(v) impacts on the employee at home as well as other staff in the office;

(vi) customer service delivery; and/or

(vii) the ability to attend meetings.

In some cases, following evaluation, it may be necessary to modify the telecommuting agreement, while in others the arrangement may be ended.

## Liabilities Associated with Working from Home

Although telecommuting is not explicitly dealt with in either the *Workplace Safety and Insurance Act, 1997*[1] ("WSIA") or the *Occupational Health and Safety Act*[2] ("OHSA"), for the purposes of WSIA and OHSA, it is likely that the home office would be considered to be the worksite. As such, employers will have obligations pursuant to both Acts with respect to their employees.

The definition of worker in the OHSA is, "a person who performs work or supplies services for monetary compensation ...." The workplace is defined as, "any land, premises, location or thing at, upon, in or near which a worker works." These definitions are broad enough to bring the telecommuter under the rubric of the OHSA.

A telecommuter also fits into the definition of worker under the WSIA. The key provision under the WSIA is section 13, which provides that a worker is entitled to benefits if the injury was a result of an "accident arising out of and in the course of his or her employment." The Workplace Safety Insurance Board (the "Board") does not have a written policy regarding telecommuting or working from home; however, it is the Board's position that if the employer is a Schedule 1 employer, and the employee is injured while at home working for the employer, that would be considered in the course of employment and an employee would be entitled to benefits.

It is reasonable and prudent for employers to seek to limit liability for injuries sustained by non-employees that occur at the home office. Limiting liability may be difficult to achieve if the employer requires third parties to attend the home office; therefore, it is recommended that the telecommuting policy prohibit any business meetings at the home office. As a practical matter, in the situation where an employee is in quarantine, meetings with third parties are unlikely. Employers in this situation should ensure through written notices that such meetings, however unlikely, are strictly prohibited.

---

1 S.O. 1997, c. 16, Sch. A

2 R.S.O. 1990, c. O.1.

◆

# CHAPTER 10 — TRAVEL POLICIES

When dealing with a public health emergency such as SARS, it may be necessary for an employer to modify its travel policies. It is not recommended that companies force anyone to travel to a country where SARS is a widespread health risk. The same would apply to countries where there are other serious wide spread health risks. Locations that have been placed under "advisory warning" status by the World Health Organization ("WHO"), the Centre for Disease Control ("CDC") or by Health Canada should be avoided if at all possible.

Whenever possible, business travel to regions where there are travel advisories should be re-scheduled. If meetings are required, employers should investigate the availability of conference call and videoconference technology as an alternative to having employees travel. In many cases this option will ensure that employees are not exposed to unnecessary health risks, while the company continues to conduct business in a cost effective manner.

There are obviously some situations where travel is unavoidable. In situations where travel to a region where there is a health risk is unavoidable, employees should be given the option to travel or not. Employees should be provided with current information regarding the possible health risks and the required precautions to assist employees to make the decision to travel or not. Employees also should be advised that there would be no repercussions if they decide not to travel. The risks associated with forcing an employee to travel, or disciplining an employee for refusing to travel, discussed in detail below. If an employee is on business in an "at-risk" country or location, the company should make every effort to bring him or her home promptly regardless of the inconvenience or expense.

There are also issues with respect to insurance coverage for employees who travel to high-risk locations. Employers should check with their insurance carrier concerning medical coverage available to employees who are travelling abroad to places where there are travel advisories. Specifically, you should ascertain the scope of hospital care, doctor's visits, private nursing care, etc., covered by the policy and ensure that the employees have all the relevant information about their health coverage. In some cases life and travel insurance policies have exclusions for situations where individuals travel to locations of civil unrest. It is also recommended that before sending employees abroad to regions where there are wide-spread health risks that it be confirmed that all applicable insurance policies will apply in the event of injury, accident or death in the foreign country.

## Liability Issues

If it is absolutely necessary to have employees travel to "at-risk" locations, it may be prudent to ask for volunteers as opposed to forcing employees to travel. An employee who refuses to accept an assignment on the basis that he or she believes it is unsafe, could take proceedings against the company if there were any kind of discipline. The Ontario Labour Relations Board allows both union and non-union employees to file a complaint of reprisal under the *Occupational Health and Safety Act*[1] and to seek full damages and reinstatement if employees feel that they have suffered adverse effects after raising a safety issue with their employer.

Because of the bar against suing for work-related injury, the only actions or legal proceedings that are likely possible would be a situation where an employee is dismissed or resigns claiming that he or she has been constructively dismissed (perhaps after returning from a trip to an at risk region or refusing to travel there). In these circumstances, it is possible that an employee would raise the fact that he or she had been put at risk for SARS and that the company had acted in a negligent or high-handed manner. The employee could then seek aggravated damages as part of a wrongful dismissal claim.

In order to reduce future liability, an employer could consider offering a travel bonus to employees who consent to travel to an at-risk area in exchange for signing a waiver that limits the liability of the company in the event that employees do become sick, terminate their employment or claim constructive dismissal. The purpose of the travel bonus is to provide consideration in order to make the waiver enforceable.

Normally, if employees get SARS in the course of their employment, either in Canada or on travel abroad, they are entitled to workers' compensation coverage, but they cannot sue the employer for work-related injury or illness. If an employee should die, his or her relatives would be similarly precluded from suing for damages arising out of the employee's death.

---

1 R.S.O. 1990, c. O.1.

It is important to understand that the company may still face civil liability if the company were negligent and third parties, other than employees, became infected as a result. For example, if the company hosted a meeting knowing that there was a risk that attendees had SARS, and a third party were in attendance, there could be liability if the third party were to become infected. It is also possible that an employee's relatives or friends could sue the company for an act or omission that led to their own illness. In the vast majority of circumstances, the company would not be liable, as the litigant would have to show that company was negligent in failing to take reasonable steps to prevent the transmission of the disease. Even so, the chance of liability only increases with ineffectiveness or inaction by the company.

◆

# CHAPTER 11 — ISSUES SPECIFIC TO AIR TRAVEL WORKERS

Employers in the air travel industry, especially airlines, will need to ensure the highest standards with respect to health and safety when faced with public health issues like SARS. Because air travel employees are interacting very closely with the public from around the world and are travelling globally, it is critical that air travel workers be educated regarding possible health risks in the countries the company flies to.

Further, it may be necessary to provide assurances to customs and immigration officials in Canada, and abroad that employees are not infected with SARS or some other infectious disease. Where it is necessary for employees to provide medical clearance before being permitted entry or exit from a particular country, it is appropriate for the employer to require the employee to provide a medical certificate confirming fitness for work or travel.

However, employers should be cautioned to avoid potential human rights violations by singling out certain groups of employees. Since federally regulated employers employ most air travel workers, the *Canadian Human Rights Act*[1] applies to their employment. The prohibition on discrimination is essentially the same as that discussed under the Ontario *Human Rights Code* in Chapter 3. Therefore, requiring medical certificates only from employees with certain ethnic backgrounds could be a discriminatory act. If medical certificates are requested they should be requested from all employees travelling to or from particular at risk countries, or from all employees who have had potential exposure to other employees who were travelling in at risk countries or exhibiting symptoms of SARS.

Given medical evidence that SARS is spread through droplets and body fluids, there is some risk that SARS could be spread by a person to another person while in the confined quarters of an aircraft. Since employees are working very close to co-workers and members of the public, it is critical that employees be educated about the symptoms of SARS so they can quickly identify whether or not they have the symptoms.

Like other workers, air travel workers have the ability to refuse work if they believe a condition creates a danger in the workplace. For federally

---

1 R.S.C. 1985, c. H-6.

regulated employers, the work refusal provisions are contained in Part II of the *Canada Labour Code.*[2] Arguably, the existence of SARS in the workplace is a "danger" as it can be an existing or potential hazard. Therefore, some employees may not want to continue working in an aircraft if they believe a co-worker or passenger has SARS. However, section 128(3) and (4) does limit the right to refuse work when working on an operational aircraft or ship. When working on an operational aircraft the worker is to report the existence of what is believed to be a dangerous condition to the person in charge of the aircraft. The person responsible for the operation of the aircraft has the authority to decide whether or not the employee is to continue working in the circumstances. The employee cannot refuse to work unless the person responsible for the operation of the aircraft approves the request. The incident is to be reported to the employer without delay, and the employer is then required to conduct an investigation.

Since air travel workers are working in such close proximity to the public, and may be exposed to individuals travelling from at risk regions, it is recommended that air travel employers establish a screening protocol to monitor employees. The screening can take place at the beginning and end of each shift, and consist of a questionnaire to be completed by each employee. The questions will allow the employer to gather information regarding whether or not employees have any SARS symptoms, or have been travelling to any of the at risk locations. This information can then be the basis for determining whether or not an employee should be advised to contact public health officials and go into voluntary quarantine. Moreover, the questionnaire will justify the employer's decision to ban an employee from the workplace until receiving appropriate medical clearance.

2 R.S.C. 1985, c. L-2, s. 128.

◆

# CHAPTER 12 — ISSUES SPECIFIC TO HEALTH CARE WORKERS

One of the key messages of this book is that all employers have an obligation to the community that we all inhabit to try to limit the damage that can be caused by a public health crisis. All employers have a responsibility to society. Some employers carry a much greater share of the responsibility than others. Nowhere is this more obvious than in the health care sector, which includes hospitals, nursing homes, homes for senior citizens, and home-care professionals. Many of the issues dealt with in health care institutions are also faced by other institutions that house and feed people, including correctional facilities, boarding schools and the hotel industry. That said, the health care sector itself, especially hospitals, has special challenges in the midst of a public health crisis.

On a daily basis, employees in health care facilities are exposed to sick people and have to adhere to strict infection control procedures. It is the health care workers who are at greatest risk when new diseases such as SARS appear. However, because the work performed by the health care workers is essential, most health care workers only have a limited right to refuse work. Accordingly, employers in the health care sector have to be extremely vigilant about communicating and training staff about new health related emergencies and infection control procedures.

## *Limitations on Work Refusals*

Section 43(2) of the *Occupational Health and Safety Act*[1] restricts the right of most health care workers to refuse work. Workers in hospitals, sanatoriums, nursing homes, homes for the aged, psychiatric institutions, mental health centers, rehabilitation facilities, residential group homes, facilities for persons with behavioural or emotional problems or a physical, mental, or developmental disability and ambulance services cannot refuse to work when the circumstance that may endanger the worker is inherent in the worker's normal work, or when the refusal to work would directly endanger the life, health or safety of another person.[2]

---

1 R.S.O. 1990, c. O.1.

2 *Ibid.*, s. 43(1).

In general this means that health care workers cannot refuse to work because they are exposed to patients who have an infectious disease such as SARS. However, health care workers still have the right to refuse work where the risk in question is not "inherent" in the job. Therefore, health care workers likely will be able to refuse work where the worker cannot be protected adequately by utilizing proper infection control procedures.

### *Personal Protective Equipment*

The Ministry of Labour has provided the following example of when a worker excluded by section 43(2) of the *Occupational Health and Safety Act* could legitimately refuse work:[3]

> An experienced medical lab technologist could not, in the course of his or her regular work, refuse to handle a blood sample from a patient with an infectious disease.
>
> But the technologist could refuse to test for a highly infectious virus where proper protective clothing and safety equipment are not available.

Accordingly, in the case of a health care worker treating a SARS patient, the Ministry of Labour may support a work refusal by a health care worker where he or she was not provided with the appropriate personal protective equipment. It is therefore critical that employers in the health care field keep abreast of proper infection control measures and ensure that all workers are provided with the recommended personal protective equipment. Not only must the personal protective equipment be provided, but also it must fit properly and the worker must be trained to use it properly.

The use of personal protective equipment in the health care field is governed by the *Health Care and Residential Facilities Regulation*[4] (the "Health Care Regulation") under the *Occupational Health and Safety Act*. Section 10(1) of the Health Care Regulation provides that when a worker is required to wear personal protective equipment, the employee must receive training and instruction on the proper use, care, and limitations of the equipments. There is a further requirement under section 10(2) that the personal protective equipment be a proper fit. Obviously, personal protec-

---

3 Ministry of Labour, *A Guide to the Occupational Health and Safety Act*, Chapter 7, revised June 2002.

4 O. Reg. 67/93.

tive equipment that does not fit properly puts employees at risk of infection.

One of the factors attributed to the spread of SARS in the health care setting was the improper use of personal protective equipment. In May 2003, the United States Centers for Disease Control ("CDC") was invited to Toronto hospitals to review the infection control protocol in place. In a news release dated May 15, 2003[5] the Ministry of Health and Long-Term Care announced that it supported the recommendations made by the CDC. Recommendations included training healthcare workers in the use and removal of personal protective equipment. As a result of CDC recommendations, the Ministry implemented guidelines on the use of masks, including fitting them, wearing them properly, inspecting them, and knowing how long to use them.

### *Infection Control Procedures*

It is important that employers in the health care field are proactive and vigilant about following infection control procedures, as well as working to improve infection control procedures. At a minimum, procedures issued by Health Canada and the Ministry of Health and Long-Term Care should be followed. The communication and enforcement of infection control procedures is essential given the potential risk to employees.

Tragically there were many front line health care workers who became ill or were quarantined as a result of exposure or possible exposure to SARS in the workplace. For those who became ill as a result of exposure to SARS in the workplace, the illness would be considered an occupational disease. In fact, directives from the Ministry of Labour advised hospitals that cases of probable SARS contracted through the workplace are to be treated as a critical injury, and cases of suspected SARS contracted through the workplace are to be treated as an occupational illness.

### *Designation, Investigation and Reporting*

The designation of probable and suspected SARS cases as critical injuries and occupational illnesses places additional obligations on the employer under the *Occupational Health and Safety Act.* Although the obligations apply to any employer where SARS is contracted through workplace exposure, so far it has been the health care employers who have borne the brunt

5 http://www.health.gov.on.ca [accessed October 9, 2003].

of the risk of workplace exposure. As the definition of what constituted a probable and suspected case changed throughout the SARS outbreak, and may continue to change as we learn more about the disease, an employer needs to keep abreast of the definition in order to know what reporting is required under the *Occupational Health and Safety Act*.

A critical injury is defined in section 1(a) of Regulation 834 under the *Occupational Health and Safety Act* as an injury of a serious nature that places life in jeopardy. Where a worker is killed or critically injured from any cause at a workplace, section 51 of the *Occupational Health and Safety Act* requires the employer to immediately notify an Inspector from the Ministry of Labour, the Joint Health and Safety Committee or Representative, and the union if any. The employer then has to notify the Minister of Labour in writing within 48 hours. In a health care facility, the contents of the notice to the Ministry of Labour must comply with section 5 of the Health Care Regulation. Essentially the notice is required to set out the particulars regarding the injured worker, the details of the injury, when it occurred, names of witnesses, names of attending physicians, and steps taken to prevent a recurrence.

In some workplaces, particularly unionized workplaces, the Joint Health and Safety Committee may have certain members designated to investigate situations where a worker is killed or critically injured. Therefore, the Joint Health and Safety Committee may be entitled to investigate the probable SARS case and report its findings to the Committee and the employer.

When the employer is advised that a worker has an occupational illness, or that a claim for an occupational illness has been filed with the Workplace Safety Insurance Board, the employer is required by section 52(2) of the *Occupational Health and Safety Act*, to notify the Director, the Joint Health and Safety Committee or Representative, and union within four days. The content of the notice is set out in section 5(5) of the Health Care Regulation. An occupational illness is defined in section 1 of the *Occupational Health and Safety Act* as "a condition that results from exposure in a workplace to a physical, chemical or biological agent to the extent that the normal physiological mechanisms are affected and the health of the worker is impaired thereby and includes an occupational disease for which a worker is entitled to benefits under the *Workplace Safety and Insurance Act, 1997*."

### *Categorizing SARS*

Given the definition of occupational illness in the *Occupational Health and Safety Act*, it seems appropriate to categorize SARS contracted in the workplace as an occupational illness. However, it is interesting that the Ministry of Labour has directed that a probable case of SARS be treated as a critical injury. Although contracting SARS certainly places a worker's life in jeopardy, arguably it should be considered to be an illness as opposed to an injury. Even so, the Ministry of Labour takes the position that if a worker is required to be hospitalized as a result of a workplace illness or injury, it generally will be considered a critical injury.

This categorization has useful practical consequences. By designating a probable case of SARS as a critical injury, as opposed to an illness, the Ministry of Labour is notified quickly and will send in an investigator. The investigator has the power to issue orders, and after a complete investigation the Ministry of Labour may decide to lay charges if the employer was in violation of the *Occupational Health and Safety Act*. Given that there were fatalties in some workplaces as a result of SARS, it is prudent and reasonable for the Ministry of Labour to want to investigate and make recommendations to ensure workers are protected, to the greatest extent possible.

◆

# CHAPTER 13 — OTHER HEALTH RELATED ISSUES IN THE WORKPLACE

SARS has been the focus of this book, however it is neither the only public health issue currently in our society, nor is it the only public health issue that we can expect to face in our working lives. Think back to the beginning of this year, when SARS was unknown, and wonder what the future might bring. With the example of SARS and other issues discussed below, all workplaces may become better informed and better prepared.

## West Nile Virus

West Nile virus is a mosquito-borne virus, and unlike SARS is not spread by person-to-person contact. A person is infected with West Nile virus when bitten by an infected mosquito. There is also some evidence that handling a dead bird infected with the virus can infect a person. Symptoms of West Nile virus vary from mild illnesses such as "West Nile fever" to serious neurological illnesses that can be fatal. Health Canada advises that many people infected with West Nile virus have mild symptoms or no symptoms at all. If illness occurs it usually happens within 5 to 15 days of being bitten by an infected mosquito. People with weak immune systems or chronic disease are at greater risk, but people of all ages and different health status are at risk. There is currently no vaccine against West Nile virus.

Although the public at large was frightened of SARS, public health experts note that West Nile will likely be more deadly. In North America, there was no real evidence of SARS being transmitted in the community, as its transmission was primarily isolated to health care settings. Individuals can be infected by West Nile anywhere there are infected mosquitoes. According to Health Canada it is not known exactly how West Nile made its way to North America. However, the first recorded outbreak in North America occurred in New York City in 1999. Since then the disease has spread to many other parts of North America, including Ontario.

Health Canada reports that over 4000 people in North America became ill in 2002 after they were infected with West Nile virus. In Ontario there were 85 probable cases of West Nile and 318 confirmed cases in 2002. As of August 12, 2003, public health officials in Ontario had reported 18 deaths related to West Nile.[1] West Nile is still an emerging disease and it continues to spread throughout North America. As of September 5, 2003 there was only 1 confirmed case and 13 probable cases of West Nile virus in humans in Ontario for 2003. However, West Nile is reported to have spread to the prairie provinces for the first time, and Saskatchewan health officials announced it had a West Nile outbreak.

In August 2003, victims of West Nile virus initiated a law suit against the Ontario government alleging the government was negligent because it failed to implement adequate testing, implement larviciding, or implement a public education campaign. The claim alleges that the government knew since 1999 that an outbreak was likely in Ontario, but failed to take appropriate action to warn the public of the dangers of West Nile virus. Assuming this legal action proceeds to trial, or even if a settlement is made public, it could become important to employers because of the public record that may be created as to what should be known about this virus and what steps should be taken to limit its spread.

This case may simply reinforce what is already known. To a large extent, any knowledge that could come from this case is currently in the public domain. It is this knowledge that is important for employers to communicate to their employees, right now. Any action steps that could make the workplace safer should be taken promptly and effectively. The mere existence of this kind of legal action establishes that those entities in society that should be doing something, which could include employers as well as government, should be sure to do whatever they can. Failure is not only bad for society, it could create liability for those that fail.

Although employers do not have to worry about spread of the disease in the workplace in the way we have worried about SARS, every employer can take proactive steps to help prevent employees from contracting West Nile virus. In particular, employers can play a key role in communicating the symptoms and preventative measures to employees. As noted in our discussion regarding SARS, employers to a certain extent have a captive audience and can provide a valuable public service by making employees aware of the danger of West Nile virus. Such communication is a benefit to employees as it may help to keep them healthy. Employers may also indirectly benefit by avoiding attendance problems by letting employees know what to do to protect themselves and their families from contracting West Nile virus. The risk of contracting West Nile virus is greatest during

---

1 http://www.health.gov.on.ca [accessed October 9, 2003].

mosquito season, which can start as early as mid-April and end as late as October in Ontario.

Employers can also reduce the chances of mosquito bites occurring on company property by cleaning up areas on its property that are potential breeding grounds for mosquitoes. For employees who work outside in the summer, employers should be providing adequate personal protective equipment, including insect repellent. It should be remembered that an employer's duty under the *Occupational Health and Safety Act* to take every reasonable precaution to protect a worker applies to West Nile virus.

The Ministry of Health and Long-Term Care recommends that anyone with the following symptoms seek medical help as these may be early symptoms of West Nile virus or many other illnesses:

- Fever
- Muscle weakness
- Stiff neck
- Confusion
- Severe headache
- Sudden sensitivity to light
- Extreme swelling or infection at the sight of the mosquito bite

Everyone is at risk from West Nile virus, particularly those who are active outside during the summer months. The same obligations under the *Occupational Health and Safety Act* discussed in Chapter 3 are equally applicable to work environments where exposure to mosquitoes is likely. Employers have a particular responsibility to ensure that outside workers are properly protected. This means that appropriate personal protective equipment such as insect repellent should be provided. Outside workers should also be provided with guidelines regarding appropriate attire for working outside. It is generally recommended that long cotton pants and shirts be worn when working in mosquito-infested areas.

An employee who believes that the workplace is unsafe as a result of a high presence of mosquitoes can also exercise his or her right to refuse work pursuant to the *Occupational Health and Safety Act*. Accordingly, employers should be sure to clear out breeding grounds for mosquitoes around buildings and work sites and to ensure that proper screens are in place where there are open doors or windows. These precautions should be taken even where the mosquito count in the area is low. It only takes one infected mosquito to infect a human, and people should not let their guard down because there do not appear to be many mosquitoes in the area.

Like workers infected with SARS in the course of their employment, workers who are infected with West Nile virus in the course of their employment will be eligible for Workplace Safety Insurance Board benefits. The Board will assess each claim on a case-by-case basis to determine whether or not the infection actually occurred in the course of employment.

The Ministry of Health and Long-Term Care recommends the following preventative measures to limit the risk of contact with mosquitoes:

- Cover up if outside when mosquitoes are most active, which is between dusk and dawn
- Wear light coloured clothing as mosquitoes are attracted to darker more intense colours
- Use insect repellent as directed
- Clean up areas where mosquitoes are likely to breed
- Clean up and empty containers of standing water
- Drill holes in the bottoms of used containers so water can't collect
- Check ponds or swimming pools to make sure the pool's pump is circulating
- Remove water that collects on pool covers
- Turn portable wading pools over when not in use
- Check eaves and drains
- Collect and recycle yard and lawn waste such as grass cuttings and leaves
- Clear out dense shrubbery where mosquitoes like to rest
- Check window screens for holes

The Ministry and Health Canada have detailed information about West Nile virus on their websites. If you would like more detailed information please visit the following websites:

- www.health.gov.on.ca
- www.hc-sc.gc.ca/english/diseases/west_nile.html.

## Second-Hand Smoke in the Workplace

There are numerous scientific studies supporting the conclusion that second-hand smoke causes serious health problems that may lead to death.[2] Yet some workers are exposed to second-hand smoke in the workplace every day. This exposure may result in liability for the employer.

In 1990 Ontario passed the *Smoking in the Workplace Act.*[3] This legislation effectively bans smoking in enclosed workplaces under the jurisdiction of Ontario law, except in designated smoking areas. Under the *Smoking in the Workplace Act* an employer can chose to make the entire workplace smoke free, however the employer can designate up to 25% of the total floor space as a designated smoking area.[4]

The legislation requires that the employer consult with the Joint Health and Safety Committee or Representative regarding establishing a designated smoking area, but it does not require the smoking area to be enclosed or separately ventilated. As a result, some workers protected by the *Smoking in the Workplace Act* may still be subjected to second-hand smoke. The legislation does require the employer to make every reasonable effort to accommodate a request from an employee to work in a place separate from a designated smoking area. Also, workers are protected from reprisal for exercising their rights under the legislation, or seeking to enforce the legislation, and have recourse to the complaint procedure under the *Occupational Health and Safety Act.*[5] An individual convicted of an offence under the legislation may be fined a maximum of $500, whereas an employer may be fined a maximum of $25,000.

*The Smoking in the Workplace Act* is not applicable to workplaces used primarily by the public. This means that bars, restaurants, and shopping malls are not required to comply with the legislation. The *Tobacco Control Act, 1994*[6] restricts or prohibits smoking in indoor places used by the public such as provincial government office buildings, schools, health care facilities and amusement arcades. However, the Regulation under the *Tobacco Control Act* specifically excludes casinos and places with a liquor

---

2 See studies referred to in W. Hyman, "Environmental Tobacco Smoke in the Workplace: The Legal Implications of Federal and Ontario Occupational Health and Safety Legislation" (1990) 4 Health L.J. 221-257.

3 R.S.O. 1990, c. S.13.

4 Early indications from the new Liberal government in Ontario (October 2003) are that smoking in the workplace will be phased out entirely within three years.

5 *Ibid.*, s. 8(1) and (2).

6 S.O. 1994, c. 10, s. 9.

license.[7] This means that there are many workplaces not subject to any legislation regarding smoking in the workplace.

However, in addition to provincial legislation, employers may have to comply with municipal by-laws prohibiting smoking in locations not covered by the legislation. For example, some municipalities have banned smoking in bars and restaurants.

Recent developments in workers' compensation adjudication are proving that it can be very costly for employers to permit smoking in the workplace. In October 2002, the Workplace Safety Insurance Board ("WSIB") granted benefits to a career waitress who was diagnosed with lung cancer. This decision was made at the first level of decision-making within the WSIB, and was made rather quickly considering it was the first decision regarding second-hand smoke as a workplace hazard.

The worker, a non-smoker, had spent decades working in smoke filled bars and restaurants. The worker had no personal direct or indirect exposure to tobacco smoke, except at work. The WSIB adjudicator determined that there was a clear causal connection between the illness and exposure to second-hand smoke in the workplace and awarded the worker benefits. The cost of the benefits paid to the worker will be transferred back to her employer and the service industry.

Also, in a 2003 Workplace Safety Insurance Appeals Tribunal (the "Tribunal") decision,[8] the Tribunal awarded damages to a worker who was forced to leave his employment as a bartender because the smoky environment caused him to have difficulty breathing. The employer had installed smoke eaters near the bar area to try and reduce the smoke, but the worker's condition did not improve. Finally his physician advised him that he had to find work in an environment without smoke. The worker used 100 days of banked sick time then quit his job to work at a lower wage rate in a smoke-free work environment. The Tribunal concluded that the worker was not permanently impaired as a result of the second-hand smoke, but the smoke was definitely an irritant making it impossible for him to continue working. The worker was awarded a wage differential until his wages reached parity with the bartending job he had to give up.

These recent cases are a warning to employers that permitting employees to be exposed to second hand smoke in the workplace may result in employees being entitled to workers' compensation benefits under the

7 O. Reg. 613/94, s. 5.

8 WSIAT Decision No. 198/03 (Josefo, Younge, Briggs), January 28, 2003.

*Workplace Safety and Insurance Act.*[9] In determining entitlement to benefits, the WSIB will consider whether exposure in the workplace could be the predominant cause of the illness, and whether or not the worker had any personal exposure to second-hand smoke outside the workplace that may be the predominant cause of the illness. The cost of WSIB may become prohibitively expensive for some employers if claims for illness as a result of second-hand smoke continue to be granted.

As an employer, steps should be taken to reduce the risk of second-hand smoke in the workplace. The following measures should be considered:

- Ban smoking in the workplace
- Comply with municipal by-laws and legislation regulating smoking
- Confine any smoking to outdoor areas or well ventilated areas
- Consider engineering controls to eliminate or reduce smoke in the workplace
- Accommodate a worker's request to work in a non-smoking area

At this point it is worth the reminder that employees have the right to refuse work if they believe it is unsafe. An employee legitimately may exercise his or her right under the *Occupational Health and Safety Act* to refuse work where it is believed that exposure to second-hand smoke is hazardous. As an employer you have a duty to investigate the complaint and attempt to find a solution. If the employee continues to refuse the work it will be necessary to have a Ministry of Labour inspector conduct an investigation. Given the recent development in workers' compensation decisions it will be interesting to see whether or not the Ministry of Labour inspectors begin to treat exposure to second-hand smoke as a hazard in the workplace under the *Occupational Health and Safety Act.*

## Chemical and Environmental Sensitivities in the Workplace

Individuals with chemical or environmental sensitivities often have adverse physical and psychological reactions to chemicals in the workplace. The reactions can range from minor irritations to incapacitation for some employees depending on the level of exposure. Individuals with chemical or environmental sensitivities exhibit symptoms such as headache, migraines, nausea, and poor concentration, sore throat, mental and physical fatigue and difficulty breathing. The reaction can be triggered by many

9 S.O. 1997, c. 16, Sch. A.

substances found in the workplace including various fragrances, cleaning chemicals, photocopier and printing chemicals, dust, and smoke.

Where exposure to substances in the workplace cause an employee to have severe physical or psychological reactions and affects the employee's ability to work, the employee may be entitled to workers' compensation benefits. Even where the chemical sensitivity is not caused by exposure in the workplace the employee may be considered to have a disability, requiring the employer to comply with obligations respecting accommodation under human rights legislation.

### *Workers' Compensation*

A worker who is diagnosed with chemical or environmental sensitivity may be eligible for workers' compensation benefits if workplace exposures are a significant contributing factor in the development of the disability. In cases where a causal link is established an employee will be entitled to workers' compensation benefits. A causal link may be found as a result of a one time exposure or prolonged exposure to a particular chemical.

For example, in a 1993 Ontario Workplace Safety and Insurance Appeals Tribunal (the "Tribunal") decision[10] a hospital worker who was sprayed in the face with perfume by a co-worker was found to be two-thirds disabled by the Tribunal and awarded pension benefits. In that decision the causal link was quite clear as immediately after being sprayed in the face with the perfume the worker had an acute reaction and required medical care. After the accident the worker experienced serious allergic reactions when exposed to any environment in which perfume had been present. The allergic reactions were at times life threatening. As a result the worker could only live and work in controlled environments.

In another Ontario decision[11] the Tribunal allowed the appeal of a school secretary who claimed her disability was due to prolonged exposures to photocopier and printer fumes. However, there was no evidence that her exposure levels were above accepted levels. The worker had pre-existing allergies to substances such as traffic fumes, perfume, tobacco smoke, and indoor and outdoor pollutants. The Tribunal noted that environmental issues in the workplace could be potentially dangerous to the

---

10 WSIAT Decision No. 818/93 (Newman, Seguin, Thompson) November 29, 1993.

11 WSIAT Decision No. 899/97 (Newman, Timms, Robb) September 29, 1998.

health and safety of workers. It accepted medical evidence that even where a minor irritant triggers the event, as in the case before it, the worker may have a strong psychological response due to the inability to control his or her exposure to the irritants. The Tribunal concluded that exposure to chemicals in the workplace was at least one significant contributing factor to the worker's disability and awarded her benefits.

### *Human Rights*

There are also human rights considerations when dealing with an employee who has symptoms of chemical or environmental sensitivity because the employee who exhibits such symptoms might be considered to have a disability for the purposes of human rights legislation.

Accordingly, the employer has a duty to accommodate the disabled employee to the point of undue hardship. Appropriate accommodation may include initiatives such as introducing a scent free workplace policy, upgrading ventilation systems, offering alternative work space or permitting the employee to work from home.

The duty to accommodate an employee with environmental sensitivities was recently considered by the Federal Court of Appeal in *Hutchinson v. Canada (Minister of Environment).*[12] That case involved an employee who, soon after beginning work with the Federal Ministry of Environment, developed extreme sensitivity to odours, particularly perfume and tobacco smoke. Over an eight-year period Ms. Hutchinson took several sick leaves. Whenever she would attempt to return to work her condition worsened and she would go off on another sick leave. The Ministry took numerous steps in an effort to accommodate Ms. Hutchinson including the following: implemented a voluntary "no-scent" policy, provided sensitivity training to employees, suggested seven alternative office locations in government buildings, encouraged telework, provided an air purifier and an industrial mask.

Ms. Hutchinson was not satisfied with the alternative work locations, was not receptive to telework, and refused to wear the industrial mask. She was however willing to pay rent to work in a private office nearby. The Ministry rejected that option due to concerns about safety and liability. In 1997 the Ministry dismissed Ms. Hutchinson on the ground that she was incapable of performing the duties of her office. After investigating Ms. Hutchinson's claim that the Ministry refused to accommodate her disabil-

12 [2003] F.C.J. No. 439, 2003 FCA 133 (Fed. C.A.).

ity, the Canada Human Rights Commission (the "Commission") dismissed her complaint on the grounds that the Ministry did attempt to accommodate Ms. Hutchinson's disability.

Ms. Hutchinson applied for judicial review to the Federal Court—Trial Division and was successful as the court ruled that she had been denied procedural fairness in the course of the Commission's investigation. The Ministry successfully appealed that decision. In a unanimous decision, the Federal Court of Appeal concluded that in the circumstances the Commission could reasonably come to the conclusion that the Ministry's response to Ms. Hutchinson's circumstances was such that an inquiry into her complaint was not warranted.

The *Hutchinson* decision gives an indication of what might be considered reasonable accommodation when dealing with an employee who is affected by chemical or environmental sensitivities. An employer that refuses to take any steps to remove scents from the workplace and/or make alternative work arrangements for an employee with chemical or environmental sensitivities may not satisfy the duty to accommodate to the point of undue hardship.

Dealing with employees who have chemical or environmental sensitivities is often difficult and can become a personal issue when an employee is complaining about the scent of a particular co-worker. Co-workers may take offence to the suggestion that their personal scent is causing another harm. Others may object to an employer mandating what personal hygiene products may or may not be used. The issue of fragrance is problematic because the issue is not isolated to perfumes and cologne. Fragrances are found in many products that people use daily, such as laundry detergent, fabric softener, shampoos and conditioners, deodorants, moisturizers, and make-up. There are also many cleaning products used in a workplace that may have fragrances. As a result it is very difficult to implement a mandatory scent free environment. However, with proper education and sensitivity training it may be possible to implement a successful voluntary scent free policy. As the implementation of a scent free policy is a health and safety issue, the Joint Health and Safety Committee should be included in the development of the policy.

## Power Failures

On Thursday August 14, 2003 at approximately 4:10 p.m., Ontario and the Northeastern seaboard of the United States experienced the biggest power failure in North American history. This massive breakdown resulted in some 50 million people being without power for much of the night and, in some circumstances, for more than 24 hours. Almost all of Ontario was affected, resulting in Premier Ernie Eves declaring a province-wide state of emergency.

It was a very hot summer afternoon and rush hour was just beginning. The power failure resulted in chaos in office buildings and city streets across Ontario. People were trapped in elevators when the power suddenly went off. Thousands of people had to walk down many flights of stairs (as many as 72 flights in Canada's tallest office tower, First Canadian Place) in order to safely evacuate buildings. City streets were in chaos, as traffic lights did not function and public transit was significantly impacted.

How a massive power outage is handled is certainly a public concern. If handled poorly, it can create significant issues of health and safety for communities and for employers.

First and foremost, there has to be an effective communication strategy and the ability, even without power, to ensure that everyone in the building is contacted. On August 14, 2003, there was significant confusion for many hours for many people, especially without access to radio or television to find out the extent of the problem. Some office towers were still telling people with disabilities to remain in place hours after the power went off. On Friday morning, many employees were unsure as to the expectations of their employers and the community at large in terms of attendance at work. Communications is clearly an area that requires significant improvement. Many employers have to do much better than they did on August 14, 2003.

It is of the utmost importance that employers consider how it will evacuate its disabled employees when there is no air-conditioning and restricted elevator service. For most people, a quick walk down several flights of stairs on a hot afternoon is little more than an inconvenience. However, for people in poor health or with a physical disability, walking down stairs is at least a very difficult task and could create serious health and safety risks. A plan must be put in place to ensure that those people who need assistance are taken care of in exiting the building. For many employers in office towers, this means working with a landlord to make significant improvements based on the lessons learned in August 2003.

Another matter to consider is what an employer should do to ensure its employees get home safely when an emergency situation results in the evacuation of the workplace, especially those employees who have special needs. On August 14, 2003, hundreds of thousands of people who rely on public transportation to get home were stranded, especially in Toronto.

Without electricity, subway trains do not operate, streetcars stop dead in their tracks, and switches on commuter train tracks are inoperable. Cellphone networks are overloaded, public telephones are severely restricted and regular telephones do not work if they are dependent on an electrical supply, which is normal for most business networks. Does an employer's obligation to its employees cease as soon as the employee is out of the building? We suggest that an employer's obligation should extend to ensure, at least, that employees with special needs get safely home.

Even if there is no legal obligation, most employers will accept that there may be at least a moral obligation to ensure that employees who are disabled receive assistance getting home, and have a support network available during the crisis. With the benefit of our recent collective experience, it seems obvious that an employer's moral duty may not be satisfied if it ensures that an employee in a wheelchair makes it out of the building only to be left stranded outside in heat, humidity, rain or snow. In our view, there are also clear issues of legal liability if such employees are left at-risk, at least in terms of health and safety and workers' compensation legislation.

In light of the recent blackout, it is recommended that every employer take a look at how the organization responded and take corrective action to ensure what went wrong this time does not go wrong next time.

◆

# CONCLUSION — REDUCING RISK IN A COMPLEX WORLD

Full-time employees spend more of their waking hours in a work setting than in any other setting, including their homes. Employers are in a unique position to communicate with employees and to educate them about issues that affect the common good of all those who are active in our society. It is also in the self-interest of employers to educate themselves and their employees about these "big issues". The failure to respond promptly and effectively is a failure to comply with the basic expectations of health and safety and human rights legislation, as well as the basic expectations of society itself. The result can be a significant decline in morale, a major increase in absenteeism and a potentially crippling effect on the ability of businesses and other organizations to operate. Even more, the result can be a crisis throughout our community that runs deeper and longer than it should.

Like SARS and the August 2003 power outage, public health issues or related crises may arrive unexpected and unannounced. Employers that are unprepared and ineffective will greatly contribute to a dysfunctional community, which is bad for all of us, and certainly bad for business. There are significant legal issues for employers. Even more important, we suggest that there is a fundamental responsibility for employers as primary participants in modern society to get ready in advance and to provide leadership when it is most required.

# Index

**Absenteeism, managing** *See also* **Human resources issues**
absenteeism policy, 50-51
accommodation, human rights, 51-52
  information relevant to accommodation, 51-52
    active steps by employers to obtain information, 52
    information to be provided to employer, 51-52
    questionnaire to be completed by medical practitioner, 52
attendance management, 50
blameless or innocent absenteeism, 49, 53-54
  dismissal for, 53-54
  long-term debilitating illnesses, 49
  non-disciplinary measures by management, 50
  short-term absences, repetitive, 49
co-operation, problems with, 52-53
  communication by employer not to constitute harassment, 52-53
culpable absenteeism, 49
  examples, 49
documentation, proper, 53
indefinite absenteeism, 54
  two-step dismissal process, 54
quarantine, absences due to, 56-57
  quarantine order under HPPA, 56
  self-quarantine, criteria for, 57
    not meeting criteria but having fear of contracting SARS, 57
    validating employee's decision, 57
recurring absenteeism, 54-56
  accommodation options, exploring all, 55, 56
  case law suggesting tolerance of excessive absenteeism, 55-56
  intermittent absenteeism potentially creating undue hardship, 56
  three-step dismissal process, 55
reporting to work when sick, consequences of, 57-58
  breaking quarantine amounting to violation of OHSA, 58
  breaking quarantine resulting in discipline, 58
  Hewlett-Packard case, 58
    quarantined employee choosing to report to work, 58
  message that should be sent to employees, 58
return to work issues, 59-60
  independent medical examination if employee agreeing, 59
  physician other than employee's, no right to compel examination by, 59
    exception under WSIA or OHSA, 59

presumption that unfit after period of time, 59-60
employer entitled to medical certificate regarding fitness, 60
further information where certificate ambiguous, 60
employer's right to determine whether employee fit, 59-60
private medical records, no right to, 59
rejection of medical certificate, providing reasons for, 60
right to know whether employee fit to perform duties, 59
sick plan, appropriate, 58-59

**Air travel workers** *See also* **Travel policies**
*Canadian Human Rights Act* applying to federally regulated employers, 79
discriminatory to require medical certificates from certain ethnic groups, 79
medical clearance to enter or exit country, 79
medical certificate confirming fitness to work or travel, 79
risk of SARS spreading in confined quarters of aircraft, 79
work refusal, 79-80
*Canada Labour Code* applying, 80
existence of SARS in workplace being "danger", 80
limits on right to refuse work on operational aircraft, 80
reporting of dangerous situation, and deciding whether to continue working, 80
screening protocol to monitor employees, 80
gathering information whether SARS symptoms, 80

**Chemical and environmental sensitivities in workplace**
human rights, 95-96
duty to accommodate, 95
*Hutchinson* decision, 95-96
accommodation options, 95-96
duty to accommodate, scope of, 96
sensitivity to odours, 95
voluntary "no-scent" policy, 95
scent free policy, voluntary, 96
sensitivity training, 96
introduction, 93-94
workers' compensation, 94-95
allergic reaction to perfume spray, 94
exposure to photocopier and printer fumes, 94
whether exposure being contributing factor to worker's disability, 94-95

**Compensation issues** *See also* **Employment insurance**
employment insurance benefits, 62
home, working from, 62
leave of absence without pay, 62
paid leave of absence or use of vacation, 61
quarantined workplace, compensation issues for, 62-63

*Employment Standards Act*, 62-63
short-term disability benefits, 61

**Contingency planning for emergencies**
affordability of contingency plan, 65
business interruption cost outweighing cost of contingency plan, 65
human resources department playing key role in developing plan, 65
issues to be considered in contingency plan, 67-68
key positions identified and back up plans implemented, 66
long-term impact of decisions considered, 67
impact on collective agreement, 67
management team developing plan, 65
operation of business without full staff, 65-66
proactive approach for better response to business interruption, 66-67
SARS being first modern, widespread medical emergency, 66
staff skills profile for filling in for absent employees, 66
worse-case scenario, plan for each, 66

**Criminal issues** *See* **Quasi-criminal issues**

**Emergency leave** *See also* **Employment legislation, application of**, employment standards
*SARS Assistance and Recovery Strategy Act, 2003*, 41-43
circumstances under which employee entitled, 41-42
physician or public health official to be contacted, 42
direction that employee not report to work, 42
retroactive to March 26, 2003, 41
SARS leave in addition to ESA leave, 41
seniority, benefits etc. continuing, 43
small employers, applicable to, 42
temporary measure, 41

**Employment insurance**
*Employment Insurance Act*, 43-45
SARS Grant Initiative for health care workers, 44-45
documentation required, 45
eligibility for special program, 44-45
maximum payment, 44
recognized health care setting, 44
weekly benefits, 44
sickness benefits, 43-44
eligibility for employees sick or quarantined, 43
medical certificate not required where quarantine order, 43-44
temporary measures, 44
two-week waiting period, waiver of, 44

**Employment legislation, application of**

employment standards, 28-31
communication required for emergency leave, 29-31
discipline not permitted where leave justified, 30
proof that eligible for emergency leave, 29, 30
protection afforded employees, 29-30
emergency leave generally, 28, 30
where need to stay home, 30
where symptoms of West Nile, SARS or other illness, 30
family members for emergency leave purposes, definition of, 28-29
"dependant", no statutory definition of, 29
"relative", meaning of, 29
reprisal provisions for disciplinary action, 31
remedies for offended employee, 31
health and safety legislation, 14-24
defences, 23
due diligence defence, 23
"officially induced error" defence, 23
duties of supervisors, 15
duties of workers, 15-16
duty of employer to provide safe and healthy workplace, 14
duty to protect employees from contracting contagious diseases, 14
duty to report dangers and contagions in workplace, 15-16
education, training, communication and action, 14-15
active steps to prevent undue exposure to health risks, 14-15
SARS or West Nile virus, 14-15
internal responsibility system for safe and healthy workplaces, 13-14
comprehensive safety policy and procedures for dealing with unsafe incidents, 13
self-monitoring and self-improvement, 13
worker input through regular inspections and committee meetings, 13
liability and penalties under OHSA, 22
penalties, 23-24
prosecution of employer and employees, 22-23
public health emergencies not new, 16
duty to reduce risk of transmission of known contagions in workplace, 16
work refusals, 17-22
external investigation, 20-21
internal investigation, 18-20
conservative approach not recommended, 19
contagion, precautions discussed where refusal due to, 19-20
infected person, whether evidence of direct contact with, 19-20
first stage being internal investigation, 19
work carrying forward unless obvious danger, 19
objective standard at second stage, 17, 20

reprisals, 21-22
complaint to OLRB, 21
OLRB having broad remedial powers, 21
disciplinary response difficult to justify, 21-22
grievance if member of bargaining unit, 21
Part VI of OHSA prohibiting, 21
right to refuse work not applying to certain sectors, 17-18
subjective right at outset, 17, 18
human rights and public health, 24-28
accommodate, duty to, 26
denial of access to workplace, 25
human rights complaint, dealing with, 26-27
complaint not substantiated, where, 27
complaint process, 27-28
corrective action where allegations substantiated, 27
training and disciplinary action, 27
duty to investigate, 27
letter to alleged harasser, 27
poisoned workplace, avoiding, 26
prohibition against discrimination and harassment based on disability, 24-25
incorrect perception unjustly stigmatizing group of people, 25
person perceived to have contagious disease protected, 24-25
person with contagious disease considered disabled, 24
privacy, 33-36
access to employee medical information limited to need-to-know basis, 36
disclose name of quarantined employee, whether to, 34
*Personal Information Protection and Electronic Documents Act* (PIPEDA), 33-35
consent to collection, use or disclosure of personal health information, 35
disclosure of personal information only to extent reasonably necessary, 35-36
contacting public health officials, 35-36
inapplicable to employment situations, 34
personal health information not to be disclosed to third parties, 34
use and disclosure of private information, without consent, if emergency, 35
where SARS or highly contagious diseases, 35
SARS prevention, 16-17
obligations varying among industries, 17
recommended safeguards, 16-17
safeguards equally applicable to other contagious diseases, 17
workers' compensation, 31-33
generally, 31
obligations of employer under WSIA, 32
occupational diseases, 32-33
compensation where impaired by occupational disease, 32

definition encompassing disease such as West Nile or SARS, 32
health care sector at risk of increased compensation claims, 33
workers with SARS symptoms having claims approved, 33
quarantined employees only covered if symptoms of SARS, 32-33
sickness required, 32
West Nile, infected with, 33
entitled to benefits if in course of employment, 33
evidence that mosquito bite in workplace, 33
qualifying for WSIB benefits, 31

**Health and safety legislation** *See* **Employment legislation, application of**

**Health care workers**
designation, investigation and reporting of SARS cases, 83-84
critically injured or killed worker, 84
investigation by Joint Health and Safety Committee, 84
notification requirements, 84
occupational illness, claim for, 84
"occupational illness" defined, 84
SARS contracted, obligations on employer where, 83-84
definition of suspected SARS case changing throughout outbreak, 84
infection control procedures, 81, 83
introduction, 81
right to refuse work, limited, 81
strict infection control procedures, 81
personal protective equipment, 82
fitting properly, 82-83
Health Care Regulation, 82
improper use contributing to spread of SARS, 83
work refusal where not provided, 82
SARS, categorizing, 85
critical injury or occupational illness, 85
notification quicker where designation as critical injury, 85
work refusals, limitations on, 81-82

***Health Protection and Promotion Act* (HPPA)**, 37-39, 56

**Home, working from**
accommodate, duty to, 70
benefits of working from home, 69-70
benefits for employees and employer, 69
disadvantages of telecommuting, 69
suitable candidates, 69-70
telecommuting policies and procedures, 69
compensation issues, 62
liabilities associated with working from home, 73

**Human resources issues**

message that should be sent to employees, 58
return to work issues, 59-60
independent medical examination if employee agreeing, 59
physician other than employee's, no right to compel examination by, 59
exception under WSIA or OHSA, 59
presumption that unfit after period of time, 59-60
employer entitled to medical certificate regarding fitness, 60
further information where certificate ambiguous, 60
employer's right to determine whether employee fit, 59-60
private medical records, no right to, 59
rejection of medical certificate, providing reasons for, 60
right to know whether employee fit to perform duties, 59
sick plan, appropriate, 58-59
maintaining wages in quarantine situations, 59
communication, 47-49
contact list, 48
information, 47-48
leadership demonstrated, 47
methods of communication, 48
unions, dealing with, 49

**Human rights and public health** *See also* **Employment legislation, application of**
accommodate, duty to, 26
denial of access to workplace, 25
human rights complaint, dealing with, 26-27
complaint not substantiated, where, 27
complaint process, 27-28
corrective action where allegations substantiated, 27
training and disciplinary action, 27
duty to investigate, 27
letter to alleged harasser, 27
poisoned workplace, avoiding, 26
prohibition against discrimination and harassment based on disability, 24-25
incorrect perception unjustly stigmatizing group of people, 25
person perceived to have contagious disease protected, 24-25
person with contagious disease considered disabled, 24

**Moral and legal responsibility of employers** *See also* **Obligations of employers**
active engagement by senior management, iii
communicable diseases in workplace, 1
past experience being limited to isolated situations, 1
emergence of new infectious diseases being normal fact of life, iii
employers having fundamental role to play in society, 2

both legal and moral roles, 2
global markets and global exchanges of infectious diseases, iii
medical and non-medical emergencies having devastating effects on enterprises, v
Severe Acute Respiratory Syndrome (SARS), 1-3
lessons learned from outbreak, 2-3
provincial emergency in Ontario, 1-2
SARS being tragic but not catastrophic, 1-2
SARS crisis catching employers unprepared, v, 2-3
need for review of policies and procedures to ensure preparedness, 2-3
SARS crisis leading to certain approaches and strategies, v

**Obligations of employers** *See also* **Moral and legal responsibility of employers**
introduction, 9
acting in best interests of organization and employees and in lawful manner, 9
legal obligations, 9-10
accommodating disability and duty not to disclose medical information, 10
employment standards and absence of employees, 10
health and safety legislation and safe workplace, 9-10
moral obligations, 10-11
communicator role being important in preventative measures, 10-11
advising employees to stay home if symptoms of disease or illness, 11
reminding employees even when no immediate threat, 11
organization structured to keep current with public health developments, 10
proactive steps to prevent employees from contracting various diseases, 10
reducing risk in complex world, 99

**Power failures**
communication strategy, 97
disabled employees, evacuation of, 97, 98
massive power outage in 2003, 96-97
transportation needs of employees, 97-98
disabled employees, assistance for, 98
public transportation, unavailability of, 98

**Privacy** *See also* **Employment legislation, application of**
access to employee medical information limited to need-to-know basis, 36
disclose name of quarantined employee, whether to, 34
*Personal Information Protection and Electronic Documents Act* (PIPEDA), 33-35
consent to collection, use or disclosure of personal health information, 35
disclosure of personal information only to extent reasonably necessary, 35-36
contacting public health officials, 35-36
inapplicable to employment situations, 34
personal health information not to be disclosed to third parties, 34
use and disclosure of private information, without consent, if emergency, 35
where SARS or highly contagious diseases, 35

**Quarantine** *See* **Quasi-criminal issues**

**Quasi-criminal issues**
generally, 37
quarantine, 37-39
    detention, 39
    enforcement of quarantine order, 38
        director's liability, 38
        individuals and corporations, 38
    *Health Protection and Promotion Act* (HPPA), 37
        powers of medical officer of health to restrict mobility of persons, 37
            power to make orders, 37
        SARS symptoms, quarantine for those exhibiting, 5, 37
        where grounds that communicable disease existing or risk of
        outbreak, 37
reporting obligation, 38
rules for employers, 39

***SARS Assistance and Recovery Strategy Act, 2003***, 41-43, 56 *See also* **Severe Acute Respiratory Syndrome (SARS)**

**Second-hand smoke in workplace**
municipal by-laws prohibiting smoking, 92
preventative measures, 93
*Smoking in the Workplace Act*, 91
    accommodating employee's request to work in separate place, 91
        reprisals, protection from, 91
    designated smoking area, 91
    inapplicable to workplaces used primarily by public, 91-92
    smoke free workplace, 91
work refusal under OHSA, 93
WSIB benefits granted to non-smokers working in smoke filled workplace, 92-93
    whether exposure in workplace being predominant cause of illness, 93

**Severe Acute Respiratory Syndrome (SARS)** *See also* **Health care workers**
air travel workers, 79-80
    *Canada Labour Code* applying, 80
    existence of SARS in workplace being "danger", 80
    limits on right to refuse work on operational aircraft, 80
    reporting of dangerous situation, and deciding whether to continue
    working, 80
    screening protocol to monitor employees, 80
        gathering information whether SARS symptoms, 80

cause of SARS still not determined, 5
close personal contact causing spread, 5
crisis catching employers unprepared, v, 2-3
emergency leave, 41-43
    *SARS Assistance and Recovery Strategy Act, 2003*, 41-43
        circumstances under which employee entitled, 41-42
        physician or public health official to be contacted, 42
            direction that employee not report to work, 42
        retroactive to March 26, 2003, 41
        SARS leave in addition to ESA leave, 41
        seniority, benefits etc. continuing, 43
        small employers, applicable to, 42
        temporary measure, 41
employment insurance, 43-45
    *Employment Insurance Act*, 43-45
        SARS Grant Initiative for health care workers, 44-45
            documentation required, 45
            eligibility for special program, 44-45
            maximum payment, 44
            recognized health care setting, 44
            weekly benefits, 44
        sickness benefits, 43-44
            eligibility for employees sick or quarantined, 43
            medical certificate not required where quarantine order, 43-44
            temporary measures, 44
            two-week waiting period, waiver of, 44
flu pandemic being realistic threat, 6-7
    Spanish Flu pandemic of 1918, 6-7
information disseminated constantly changing, 5-6
    challenge for human resources departments, 6
preventing SARS, steps in, 16-17
    recommended safeguards, 16-17
        safeguards equally applicable to other contagious diseases, 17
provincial emergency in Ontario, 1-2
quarantine for those exhibiting symptoms, 5
resurfacing of SARS at any time, 6
symptoms of SARS, 5
widespread and unexpected public health emergency, 7

**Telecommuting** *See* **Home, working from**

**Travel policies** *See also* **Air travel workers**
"advisory warning" status for locations by World Health Organization(WHO), 75
    business travel to be re-scheduled, 75
insurance and medical coverage issues, 75-76
introduction, 75-76

liability issues, 76-77
negligence in hosting meeting where risk that attendees had SARS, 77
liability if failure to take reasonable steps to prevent transmission, 77
refusal to accept assignment, consequences of, 76
constructive dismissal claim where dismissed or resigning, 76
travel bonus for employees consenting to travel to at-risk area, 76
volunteers for "at-risk" locations, 76
workers' compensation coverage if contracted SARS, 76
option to travel or not where health risk and high-risk locations, 75
advising that no repercussions if they deciding not to travel, 75

**West Nile virus**
employer's role in communicating symptoms and preventive measures to employees, 88
legal action against Ontario government for inadequate testing and warning, 88
mosquito-borne virus, 87
North America, probable and confirmed cases in, 87
outside workers properly protected, 89
preventive measures to limit risk of contact with mosquitoes, 90
proactive steps to protect employees and their families, 88-89
breeding grounds for mosquitoes to be cleared out, 89
refusal to work pursuant to OHSA, 89
symptoms of virus, 87, 89
medical help recommended where certain symptoms, 89
WSIB benefits where infected at work, 33, 90

**Work refusals** *See also* **Employment legislation, application of**, health and safety legislation
air travel workers, 79-80
*Canada Labour Code* applying, 80
existence of SARS in workplace being "danger", 80
limits on right to refuse work on operational aircraft, 80
reporting of dangerous situation, and deciding whether to continue working, 80
screening protocol to monitor employees, 80
gathering information whether SARS symptoms, 80
external investigation, 20-21
health care workers, 81-82
work refusals, limitations on, 81-82
internal investigation, 18-20
conservative approach not recommended, 19
contagion, precautions discussed where refusal due to, 19-20
infected person, whether evidence of direct contact with, 19-20
first stage being internal investigation, 19
work carrying forward unless obvious danger, 19

objective standard at second stage, 17, 20
reprisals, 21-22
complaint to OLRB, 21
OLRB having broad remedial powers, 21
disciplinary response difficult to justify, 21-22
grievance if member of bargaining unit, 21
Part VI of OHSA prohibiting, 21
right to refuse work not applying to certain sectors, 17-18
second-hand smoke in workplace, 93
work refusal under OHSA, 93
subjective right at outset, 17, 18
West Nile virus, 89
refusal to work pursuant to OHSA, 89

**Workers' compensation** *See also* **Employment legislation, application of**
chemical and environmental sensitivities in workplace, 94-95
generally, 31
obligations of employer under WSIA, 32
occupational diseases, 32-33
compensation where impaired by occupational disease, 32
definition encompassing disease such as West Nile or SARS, 32
health care sector at risk of increased compensation claims, 33
workers with SARS symptoms having claims approved, 33
quarantined employees only covered if symptoms of SARS, 32-33
sickness required, 32
West Nile, infected with, 33, 90
entitled to benefits if in course of employment, 33
evidence that mosquito bite in workplace, 33
qualifying for WSIB benefits, 31